ROYAL BLUE LINE

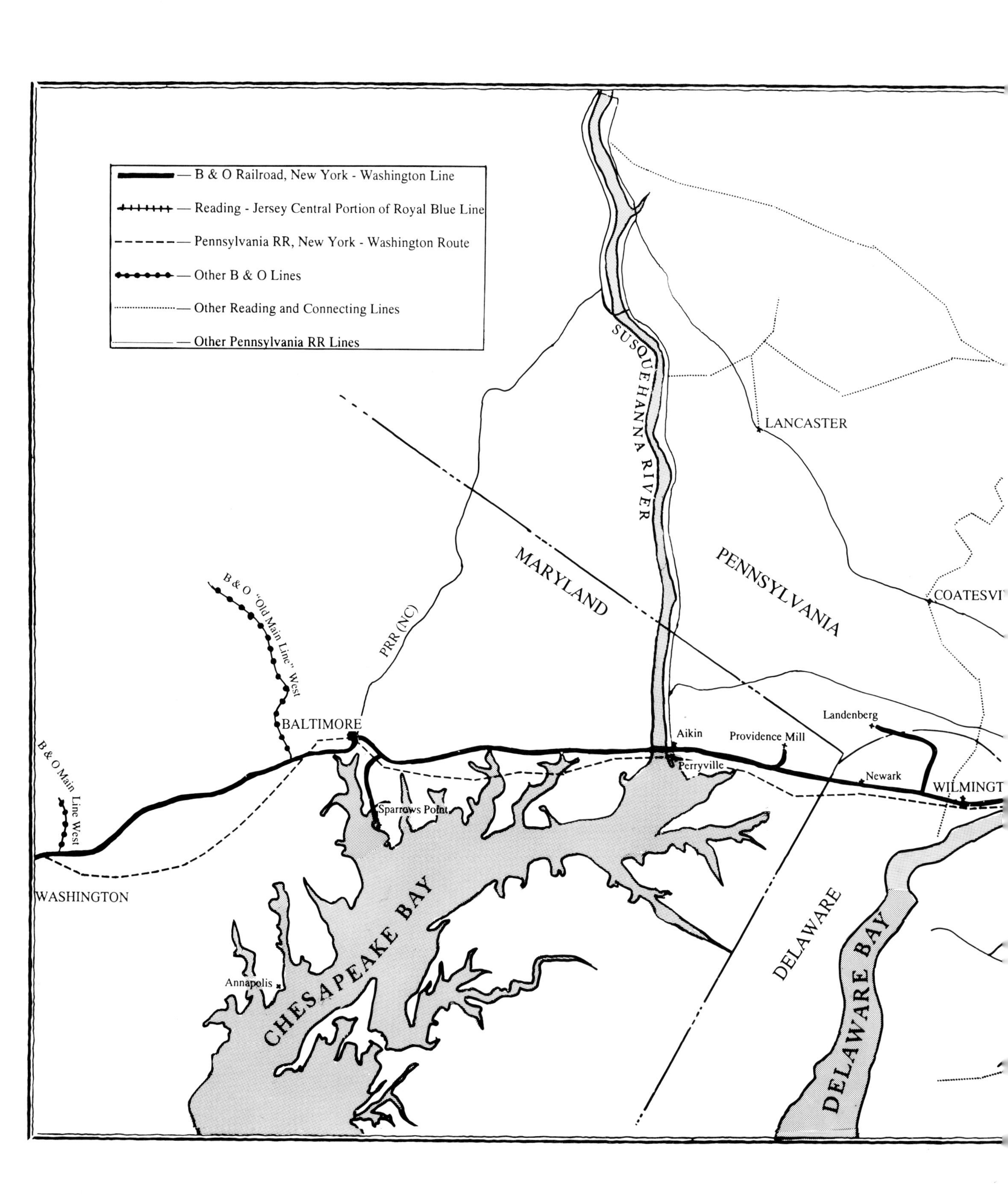

— B & O Railroad, New York - Washington Line
— Reading - Jersey Central Portion of Royal Blue Line
— Pennsylvania RR, New York - Washington Route
— Other B & O Lines
— Other Reading and Connecting Lines
— Other Pennsylvania RR Lines
SUSQUEHANNA RIVER
LANCASTER
PENNSYLVANIA
MARYLAND
COATESVI
PRR (NC)
B & O "Old Main Line" West
BALTIMORE
B & O Main Line West
Aikin
Providence Mill
Landenberg
Perryville
Newark
WILMINGT
Sparrows Point
WASHINGTON
CHESAPEAKE BAY
Annapolis
DELAWARE
DELAWARE BAY

ROYAL BLUE LINE

THE CLASSIC B & O TRAIN BETWEEN WASHINGTON AND NEW YORK

HERBERT H. HARWOOD, JR.

WITH A FOREWORD BY COURTNEY B. WILSON

THE JOHNS HOPKINS UNIVERSITY PRESS
BALTIMORE AND LONDON

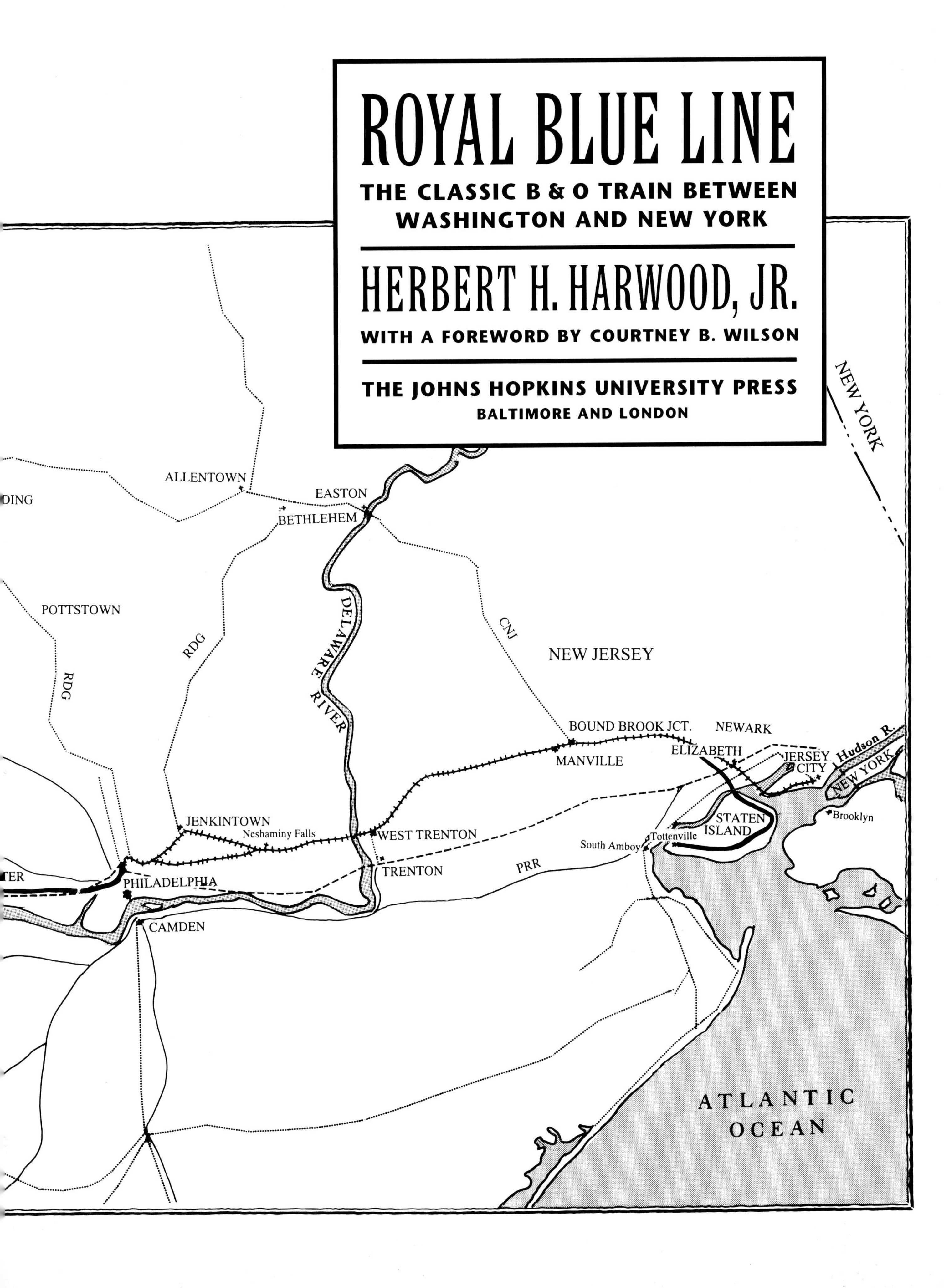

Printed in the United States of America on acid-free paper

Originally published in a hardcover edition by
Greenberg Publishing Company, Inc., 1990
Johns Hopkins Paperbacks edition, 2002
9 8 7 6 5 4 3 2 1

The Johns Hopkins University Press
2715 North Charles Street
Baltimore, Maryland 21218-4363
www.press.jhu.edu

Library of Congress Cataloging-in-Publication Data

Harwood, Herbert H.
Royal Blue Line : the classic B & O train between Washington and New York / Herbert H. Harwood, Jr.
p. cm.
1990 edition does not have subtitle.
Includes bibliographical references and index.
ISBN 0-8018-7061-5 (pbk. : alk. paper)
1. Baltimore and Ohio Railroad Company. 2. Railroads—United States.

TF25.B2 H383 2002
385'.0974—dc21 2002016048

A catalog record for this book is available from the British Library.

CONTENTS

Appendices and Other Matter

A Personal Preface

The seed of this book really was planted some fifty years ago. As a nine-year-old living a narrow suburban life not far north of New York City, my biggest adventure was a drive in our 1937 Ford to my mother's former home in Plainfield, another suburb across the river in New Jersey. The object of the trip was a visit to my Edwardian-era grandparents and a formidable collection of maiden great-aunts, portly great-uncles, and a miscellaneous assortment of distant cousins whose names and exact relationships I would promptly forget.

But not forgotten were the walks I could take to the nearby Central Railroad of New Jersey station. To Plainfielders, the railroad was always simply the "Jersey Central." It was an excitingly different place from the station in my own home town, which offered only a monotonous succession of electric suburban trains. At Plainfield I was always assured of a constant and fascinating parade of steam-powered trains from three different railroads on the four-track main line. And since Plainfield (to my own bafflement) was considered an important spot, all passenger trains would stop there for closer inspection.

Fat-boilered Mikados of the Jersey Central and the Reading Company stormed purposefully past with long merchandise and coal trains bound for the yards and piers at New York Harbor, churning up lots of dust. More common and more charming were the Jersey Central's little ten-wheeled camelbacks, which regularly chuckled in and barked out with their suburban trains. Lithe Reading Company Pacifics and their beefier Jersey Central counterparts brought in the more important trains — the New York-Philadelphia "Two Hour Flyers" and a scattered assortment of trains bound for unknown places like Allentown, Wilkes-Barre, Reading, and Harrisburg. These high-wheeled steam engines were always impressive, although their trains were drab and did not seem much different from the ordinary suburban equipment. The dazzling exception was the Reading's shining stainless steel *Crusader* with its matching streamlined shovel-nosed Pacific, the single gesture to modernism on the Philadelphia run.

Then there was the Baltimore & Ohio, which also used the Jersey Central's tracks. The B & O was a class apart. Its trains went to impossibly remote and truly important places like Washington, Chicago, and St. Louis. They exuded nobility, with stately, heavy long-distance cars, many of them streamlined, and painted a pleasing blue and gray with gold striping. As they stopped, swarms of white-jacketed porters would descend to pick up bags and hand passengers into the cars. And wonders still more, some of the B & O trains were hauled by huge streamlined

Class comes to Plainfield: Bound for New York, a shiny set of E-8 diesels rolls a typically mixed consist of the National Limited into the Jersey Central station in June 1954. H. H. Harwood, Jr. photograph.

diesels, an exciting new type of locomotive seen nowhere else in my limited territory. Regally painted in blue, gray, and black, with touches of silver and gold, they seemed supremely authoritative. Showing almost no outward effort, they glided in and rumbled away with their long, heavy strings of sleeping cars, diners, lounge cars, roomy coaches, and, on some, smooth-end observation cars carrying names like *The Royal Blue* and *The Columbian* in a fancy script. Soon they disappeared over the flat horizon, heading for all those hazy, mystical places I had never seen — and perhaps never would.

By World War II we were living in Washington and my world was becoming less mystical. But now I was actually riding those B & O trains back to Plainfield and New York for the family visits. To a young teenager the trains now seemed a bit less noble, but were solid, friendly, and comfortably homey. The scenery along the way was hardly exciting, but entering New York City never ceased to be an adventure — and a different one each time. B & O trains, you see, did not really enter New York; they ended at the sprawling Jersey Central ferry terminal in Jersey City and were met by a fleet of special buses for the trip into Manhattan.

Before the train arrived in Jersey City, our luggage would be whisked away with reassurances that it would reappear for us at the 42nd Street bus terminal next to Grand Central. The streamlined buses were waiting by the train doors, looking as regal as a bus could look. After we settled into their soft imitation leather seats, they would drive off in a convoy and onto the ferry for New York. Once aboard the boat, we would immediately abandon the bus for a place on the deck, watching lower Manhattan slowly approach us across the water. All around were tugs, barges, railroad carfloats, and ocean-going ships, and an intermittent chorus of steam whistles from everywhere. Then came the clanging bells, churning water, and intently-moving deckhands as the ferry docked at its slip and was secured, and we scurried back into the bus for the ride uptown. Thanks to the peculiarities of the undeveloped teenage mind, the most memorable landmark along the way was always the giant "S. Klein" sign at Union Square. At the clean, modern 42nd Street terminal our luggage was indeed handed to us as promised.

In the years after, we moved elsewhere (several elsewheres, in fact) and I lost touch with the B & O's New York service. Sadly, so did most other one-time train riders. I was 400 miles away and unheeding when it finally shriveled up and died. By then a wise and cynical young adult, I had realized that the nobility and style which I had seen twenty years earlier was partly a case of putting on a brave face for an operation which made no business sense and perhaps never did.

Now, thirty years later still, I have probably covered a distance equivalent to a trip to Mars and back inside jets and on Interstate highways. And in the meantime, not only is the B & O gone from New York, but also all the trains I saw at Plainfield — and all the people that I knew. There is no Jersey Central, Reading, or even B & O anywhere. But I can still clearly see those stately Pullmans vanishing into the horizon and feel the mystery.

Herbert H. Harwood, Jr.

Symbolic of the era of the growing wealth and power of the anthracite-region railroads, the Jersey Central's impressive 1874 westbound station at Plainfield served CNJ, Reading, and B & O trains. By the time this 1951 photograph was made, the power and wealth had vaporized; symbolically too, the station later was reduced to a single story. E. H. Weber photograph.

FOREWORD

Generations of East Coast travelers readily associate the name *Royal Blue* with memories of the deluxe passenger-train service that the Baltimore & Ohio Railroad once delivered between the nation's capital and New York City. *Royal Blue* conjures images of truly genteel service, streamlined design, and carefully selected blue-and-gold color schemes—sleek locomotives and luxury appointments. Herbert H. Harwood, Jr.'s *Royal Blue Line* delightfully explores the history of this legendary train.

As is true with so much American railroad history, the tale involved intrigue, politics, money, and power. Chartered in 1827, the B & O's aim was to build westward, to bring the bounty of interior America to the port on the Chesapeake. It soon served the nation's capital, as well. But in the late 1870s, at a critical point in its corporate evolution, the B & O realized that its size and power had been eclipsed by a few of its rivals in the northeast. When the B & O tried to land a place along the route we now know as the Northeast Corridor, it found itself a latecomer, an old-timer in an upstart role. The B & O had to struggle in places where other railroads had already established power and presence, and in a sense the fight to reach Manhattan never ended. The B & O line to New York came about only after a patchwork of deals between the B & O and cooperating as well as rival railroads. Except during short periods of time, B & O trains did not enter the City of New York. Instead, the railroad provided service by means of buses and ferries that operated from the western banks of the Hudson River and eventually dropped off passengers in downtown Manhattan. To help make the jury-rigged line appealing and otherwise to compete with the Pennsylvania and the New York Central, the B & O launched a brilliant marketing campaign that traded on its very age and location.

The effort began in 1880 when the B & O hired "Major" Joseph G. Pangborn as America's first railroad publicity agent. Pangborn, a western newspaperman and self-styled showman, singlehandedly transformed the public and corporate image of the B & O. Magnificent promotional materials, posters, timetables, publications, and a series of gold medal–winning exhibitions methodically turned the heads of railroad barons, their clients, and the public. Pangborn coined phrases that formed part of the railroad's lexicon for generations. Trading on the B & O's historical significance and established territory, he billed the railroad as historic, "The Picturesque Line of America," the country's "Model Fast Line." He utilized its direct connection with Washington, D.C., to bill the railroad as the principal artery to the nation's capital, from east and west alike. Advertising and packaging became a corporate trait of the B & O, unmatched by any other line; and while using a broad brush to market all of its lines, the B & O above all created the mystique of the Royal Blue.

Some of the attention the line surely deserved. Given the experimental nature of the B & O itself, the Royal Blue not surprisingly supplied a number of technological firsts: the nation's first air-conditioned trains, innovative high-speed steam locomotives and early diesels, fresh interior and exterior designs, and creative passenger accommodations. Meantime, the line became almost a matter of obsession for a string of B & O presidents. None of the railroad's other premier lines captured the level of attention the Royal Blue did; the amount of corporate pride, marketing effort, design, construction, and energy poured into this line far surpassed the margin of profit it could ever generate. Even during times of extreme economy and near bankruptcy, the autocratic presidents of the B & O regarded the Royal Blue as a symbol—not to be tarnished—of their personal pride and power.

Memories of Royal Blue trains doubtless testify to the B & O's success in marketing this difficult, expensive, and troubled passenger line to the public. As I often perceive when speaking with people who used this line in and out of Baltimore, no one cares to remember struggle, politics, or sham. Even the inconvenience of transfers from train to bus to ferry have long disappeared, leaving behind only thoughts of genteel and luxurious service, premier accommodations, magnificent Chesapeake cuisine, on-time trains, and a regal name befitting the crack train of the first American railroad.

Courtney B. Wilson, Executive Director
Baltimore & Ohio Railroad Museum
March 2002

PROLOGUE

A Winning Loser

Royal Blue — somehow the name produces a twinge of recognition even for those who never knew what it represented. It was indeed one of the most memorable images in the transportation business, an inspired blend of majesty and mystique which properly fitted its creator — the Baltimore & Ohio Railroad. First conceived in late Victorian times to promote a new railroad line which desperately needed all the promotion it could get, the image led its own mystical life. It appeared, expanded, disappeared, reappeared in another form, and again changed. Royal Blue Line...Royal Blue Trains...*The Royal Blue* all meant different things at different times. But essentially they all symbolized one thing: the B & O's regal route between New York, Philadelphia, Baltimore, and Washington.

"Royal Blue" may have been merely a marketing device, but in this case the image and the reality were unforgettably intertwined. For almost seventy years, the Royal Blue Line was B & O's showpiece. Physically, it was the best engineered, most heavily built, and elaborately equipped part of the company's 6,300-mile system. Its passenger trains always were the finest and fastest that the railroad could offer, powered by locomotives specifically designed for the service. And they remained so until the end; when it became clear that the service was no longer financially supportable, it was ended at once and *in toto*, avoiding the indignity of a lingering death. Today, over thirty years after the last Royal Blue Line passenger trains ended their runs, they are remembered with fondness and some awe.

But the Royal Blue Line was much more than just a passenger service. As B & O's main line to Philadelphia and New York Harbor, it also was a busy freight route which once included an extensive "navy" of tugs and floating equipment as well as a complex array of waterfront piers. It was a major carrier of food to Philadelphia and coal to New York.

It also was a railroad line of anomalies, unlike anything else on the B & O. Although advertised and sold as a through B & O operation, it really was not. Half of the route, the key portion between Philadelphia and New York, was never owned or directly managed by the B & O. Yet passengers carried B & O tickets and boarded at B & O stations in New York; freight was shipped in and out of New York on B & O waybills and handled through B & O facilities. Indeed, for the B & O, New York was in a railroad twilight zone, a city which the company reached but did not reach. Then there were some lesser peculiarities. For example, the Royal Blue Line achieved some kind of distinction by substituting a steam-powered heavyweight train for an almost new diesel-powered lightweight — and advertising it as a step forward. A key component of the line was a B & O subsidiary separated from the nearest B & O track by over 73 miles: it operated electric rapid transit trains, camelback steamers, and, at one time, New York's famous Staten Island ferry.

Oddities aside, the Royal Blue Line was notable for a more important reason: it was the birthplace or proving ground for some of the most significant innovations in 20th Century railroading. Its route through Baltimore was the country's first main line railroad electrification. At its Manhattan freight terminal, the B & O installed its first diesel locomotive, which was the second diesel put in service on any American railroad. The Royal Blue Line operated the

B & O's New York passengers entered the city this way — slow, perhaps, but always breathtaking. The Jersey Central's ferry Plainfield heads toward Liberty Street in 1947 with B & O buses aboard. Likely most of the bus passengers are outside on deck enjoying the view. Tracey Brooks Collection, Steamship Historical Society.

country's first regularly-used air-conditioned railroad car, its first fully air-conditioned train, the first main line passenger diesel, and one of the earliest lightweight streamliners.

Structures on the route included the world's longest vertical lift bridge, the B & O's longest tunnel (which, when built, was an engineering challenge of national interest), and its longest bridge of any type. On the route, too, were B & O's only track pans, a device which allowed steam-powered trains to take water at speed.

And to any general business historian, the B & O's New York-Washington passenger route is a case study of constantly creative marketing techniques. Beginning with the "Royal Blue" image itself, these included the design of equipment and services which fitted the image, a multitude of distinctive advertising campaigns and promotional devices, and one of the earliest examples of a precisely-coordinated rail-bus intermodal service.

Sadly, there was a dark side too. There seems to be strong evidence that the line's creation was the impulsive act of an unstable mind. Unquestionably, its construction helped to push the already-weak B & O into financial collapse and receivership. Cold-eyed critics have complained that the B & O wasted its financial resources on an expensive, ill-conceived, and unnecessary venture at a time when the money was desperately needed to improve its many physical shortcomings elsewhere.

And they are probably right. In truth, many of the Royal Blue Line's marketing and mechanical innovations were forced by a grim necessity; they were needed to overcome fundamental competitive weaknesses which never could be fully conquered. Born crippled, the line had to struggle painfully (and expensively) to build strength against a vigorous, powerful, and entrenched rival: the imperial Pennsylvania Railroad. At times it succeeded, but never for long. In the 72 years that B & O operated in the New York-Washington passenger market, it was an effective competitor for only two short periods totaling 23 years and for one of those periods it depended on a vital Pennsylvania Railroad facility. Its fight for freight business was slightly more equal, but also was hampered in various ways.

A belated birth was the Royal Blue Line's primary curse. In railroading, as in real estate, location is everything. The railroad whose tracks directly reach the industries and community centers gets most of the business. The B & O's route was not put together until 1886, which by the standards of eastern railroad history was late. It was the last large railroad to enter Philadelphia and the last into New York. By then the best locations had been taken, most of them by the Pennsylvania. Even at that, B & O was only able to complete its line as far as Philadelphia; rapidly deteriorating finances and the iron hand of J. P. Morgan prevented it from building farther east. To reach New York it had to depend on two friendly connections, the Reading and the Central Railroad of New Jersey. Although their relationship worked well enough, B & O always was limited in reaching sources of business — especially freight traffic in densely-industrialized eastern New Jersey. Furthermore, B & O never could completely control its two partners, and always had to match its interests with theirs. When the Reading and Jersey Central eventually went hopelessly bankrupt and were folded into an unfriendly Conrail, it lost its New York link altogether.

A far more famous shortcoming was its New York passenger terminal. In 1910 the Pennsylvania delivered an almost-mortal blow when it opened its Penn Station in midtown Manhattan, leaving the B & O on the wrong side of the Hudson River in Jersey City with only a ferry to New York itself. After this it was really no contest, although the B & O certainly tried. Thanks to the exigencies of World War I and some deft negotiating afterward, B & O actually used Penn Station for eight and one half years, but obviously the blessing could not last. For most of its life after 1910 the Royal Blue Line had to live on its wits and resourcefulness. That its passenger service lasted as long as it did, until 1958, is a tribute to determination, imagination, and corporate pride.

Much of that pride and determination came directly from the company's two strongest presidents: John W. Garrett in the 19th Century and Daniel Willard in the 20th. The autocratic Garrett personally created the line as his last grand gesture, a gesture of defiance and frustration as much as anything else, to show the world that the B & O was a major power. Willard, who ruled the railroad more gently but equally autocratically for 31 years, steadfastly and stubbornly kept the Royal Blue Line as the company's showpiece and, in the process, made so much railroad history.

Perhaps the pride was misplaced. But for better or worse, the B & O's Philadelphia line does survive today as a healthy, strategic all-freight route, although with a different mission and appearance from what John Garrett and Daniel Willard intended. To get to that point it followed one of the oddest careers of any railroad.

If adversity shaped the Royal Blue Line, perversity created it. A New York extension was the last thing on the minds of the B & O's founders and its early managers; in fact, the idea most likely would have been an anathema to them. And had John Garrett died slightly earlier than he did, it probably never would have existed. But the twists of railroad power politics and personal pride started a process which, once begun, had the B & O heading for the Hudson River with a heedless vengeance.

GENESIS:

A BELATED AWAKENING

1835-1881

In the beginning there was only Baltimore. Baltimore was not merely the Baltimore & Ohio Railroad's birthplace, it was its entire reason for existence. Businessmen in the Chesapeake Bay port city created the B & O in 1827 in a mixture of bravery, vision, and desperation. It was their only hope to recover an advantage in Baltimore's two-way rivalry with the ports of New York and Philadelphia for inland trade. Two years earlier, New York had scored a decisive hit when the Erie Canal was completed, giving it a far cheaper and easier route to the west than Baltimore's turnpikes and its National Road connection across the Allegheny Alleghenies.

With its turnpikes suddenly obsolete and a competitive canal impossible, Baltimore's bankers and merchants turned to the railroad — a means of transportation which was just in its infancy and, for the distances and terrain which the B & O had to conquer, completely untried. But, for the city's survival and growth, it had to be built.

The B & O's purpose was to link Baltimore with the Ohio River, the gateway to the newly-developing Midwest. After a grueling off-and-on struggle, it finally reached the river at Wheeling in December 1852, almost 26 years after the company was created. Afterwards it gradually extended farther westward through connecting lines and affiliated companies. By 1857 Baltimore was connected after a fashion with Cincinnati and St. Louis, and in 1866 the B & O took over operation of a line to Columbus, Ohio. In the meantime it also had tapped the bituminous coal fields of western Maryland and West Virginia, and greatly strengthened Baltimore's traditional role as a major bulk cargo port.

Throughout all of this, the control and financing of the B & O remained tightly within Maryland, although by then somewhat uneasily shared between Baltimore's business interests and the city and state governments. In 1858 these groups picked a new president to represent them, John W. Garrett, himself a Baltimore merchant and investment banker. Physically imposing, outgoing, and exceptionally strong-willed, Garrett ruled the railroad for the next 26 years, gradually making it into his personal fiefdom. Such was the strength of his ego and will that, according to a B & O vice president of the 1870s, "in no single instance did Mr. Garrett give [the Board of Directors'] opinions a moment's consideration when they differed from his own." Nonetheless Garrett was a Baltimorean and never forgot his primary mission; the B & O remained Baltimore's servant and savior, with its route structure and traffic flows firmly anchored there and feeding the city.

Baltimore was hardly unique. Soon enough, New York and Philadelphia also had railroad lines reaching west. By the eve of the Civil War, the Philadelphia-based Pennsylvania Railroad owned a through line from Philadelphia to Pittsburgh and, in the process, had frustrated the B & O from building into Pittsburgh. At Pittsburgh it met other railroads extending farther west. Also by that time the PRR had acquired an unusually astute and aggressive executive management, and clearly was becoming B & O's arch rival. New York had two connecting railroad companies which together formed a New York City-Buffalo line. (It was not until 1867, however, that a 73-year-old former ferryboat operator named Cornelius Vanderbilt put them together as one system and created the later New York

Early rail travelers between New York and Washington rode B & O trains like this on the Baltimore-Washington leg of their trips. This bucolic 1839 scene actually is deep inside Washington, about five blocks north of the Capitol. The engine is a William Norris 4-2-0. Library of Congress Collection.

Central.) Also connecting New York and Lake Erie was the problem-plagued Erie Railroad, terminating on the west side of the Hudson at Jersey City, New Jersey. Through their separate railroads, Baltimore, Philadelphia, and New York concentrated mostly on competing for western trade and developing their own hinterlands.

But while the three port cities looked westward with their growing trunk line railroads, they also did a brisk business with one another and with the other Atlantic coastal communities — which included Washington, DC, then not much more than an eastern frontier town with unpaved streets but, after all, the nation's capital. Together these cities formed the country's largest population concentration and, rivals or not, there was considerable traffic between them — although more heavily passenger travel than freight movement. Yet their commercial and financial competition seemed to preclude any one single railroad company from building a through line between New York and Washington.

Instead, starting in 1832 with New Jersey's Camden & Amboy Rail Road, a patchwork of independent operations appeared which, willy-nilly, gradually coalesced into a coherent "corridor." B & O itself built one of the earlier and important segments and, in fact, was the only one of the large trunk line railroads to do so. In 1835 it opened a branch between Baltimore and Washington, forty miles apart. At the time, B & O's primary interest was in building its main line west to the Ohio; the Washington branch was more the product of political sensitivities than commercial desire, and was partly financed by the state of Maryland. Nevertheless, it was an exceptionally well-engineered and solidly-built railroad for the time; in later years it could handle heavy trains at 80 mph speeds (and sometimes more) without any realignment or significant rebuilding. And although B & O considered it a sideline only marginally related to its main line business, the branch rapidly turned into a busy and profitable passenger route.

Once the Washington branch was completed, it became the southern end of a complex water-rail-water-rail-water-rail route between New York and Washington. The trip in 1835 took patience and strength, but it was a notable improvement over a stagecoach on post roads or the length-

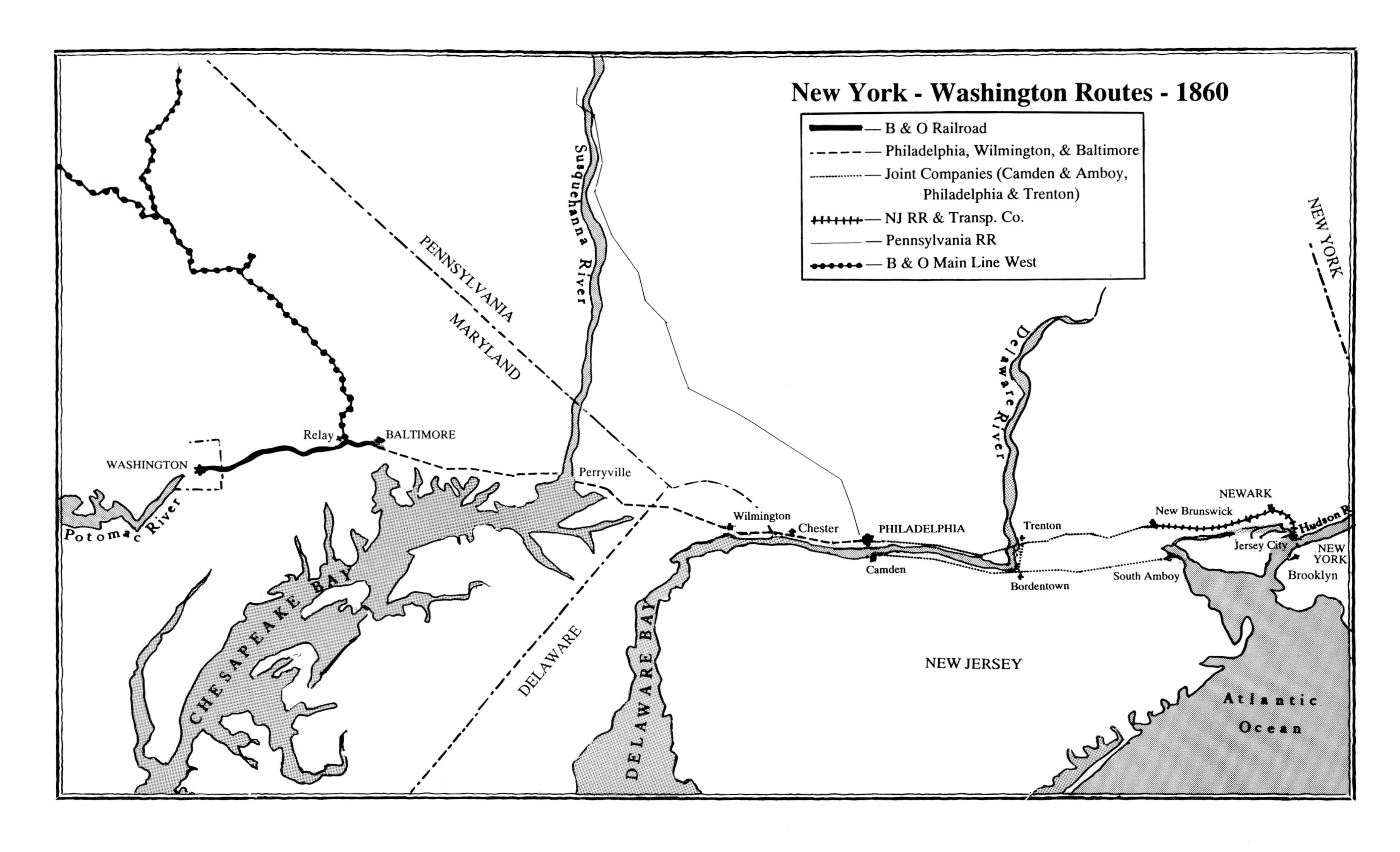

New York - Washington Routes - 1860
— B & O Railroad
— Philadelphia, Wilmington, & Baltimore
— Joint Companies (Camden & Amboy, Philadelphia & Trenton)
— NJ RR & Transp. Co.
— Pennsylvania RR
— B & O Main Line West
NEW YORK
PENNSYLVANIA
MARYLAND
DELAWARE
NEW JERSEY
Susquehanna River
Delaware River
Potomac River
Hudson R.
CHESAPEAKE BAY
DELAWARE BAY
Atlantic Ocean
WASHINGTON
Relay
BALTIMORE
Perryville
Wilmington
Chester
PHILADELPHIA
Camden
Trenton
Bordentown
South Amboy
New Brunswick
NEWARK
Jersey City
NEW YORK
Brooklyn

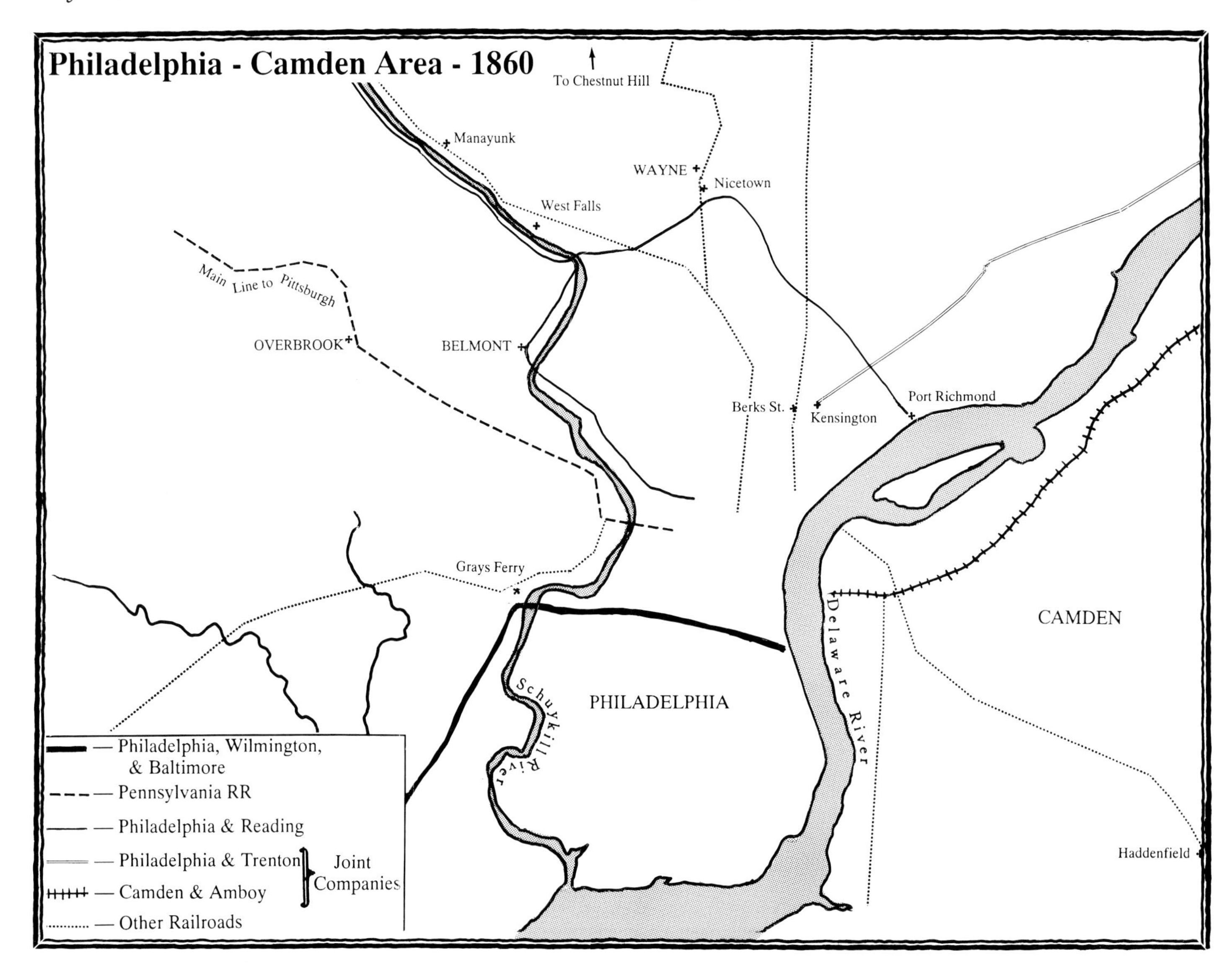

ier all-water route. The New York-Washington traveler first took a short steamer ride from Manhattan to South Amboy, New Jersey, at the south end of New York's harbor. From there the Camden & Amboy Rail Road carried him overland sixty miles to Camden, New Jersey, opposite Philadelphia on the Delaware River. Then came another steamer ride down the Delaware to New Castle, Delaware, where a New Castle & Frenchtown Rail Road train waited to haul him across the sixteen-mile neck of land separating the Delaware River and Chesapeake Bay. At the railroad's dock at Frenchtown, Maryland, a now-forgotten port on the Elk River, the tired traveler again climbed onto a steamer for the trip down the Chesapeake Bay to Baltimore. Arriving at Baltimore's "Basin" (now called the Inner Harbor), he could walk two or three blocks to the B & O station at Pratt and Charles Streets and climb onto a car for Washington. After a two-hour trip (which included being hauled by horses at each end), he was finally deposited at the B & O's unprepossessing terminal, a former boardinghouse on Pennsylvania Avenue at 2nd Street.

Not surprisingly, this tedious and unwieldy string of boat and train transfers did not last long. By 1840, less than five years after B & O finished its Washington branch, a complete chain of railroads had appeared along the New York-Washington route. It was still an arduous trip, using five separate railroad companies and requiring at least three transfers, but it was mostly all-rail. Now the New York-Washington traveler took a short ferry ride directly across the Hudson River to Jersey City, opposite lower Manhattan. From Jersey City the New Jersey Rail Road & Transportation Company ran 32 miles to New Brunswick, New Jersey where it met a Camden & Amboy line which continued on to Trenton. To get from Trenton to Philadelphia there were two choices. The Philadelphia & Trenton Railroad ran there directly, following the west side of the Delaware River and ending at a station in the remote Kensington section of Philadelphia. Or, the Camden & Amboy offered an alternative route along the east bank of the river to Camden, where riders took the Philadelphia ferry across. In the earlier years, passengers preferred the C & A route and the ferry ride from Camden, which landed them near the city center; the Philadelphia & Trenton's Kensington terminal at Front and Berks Streets was two and one half miles away over congested streets.

Washington-bound passengers left Philadelphia on the Philadelphia, Wilmington & Baltimore Railroad, the longest single line in the chain. The PW & B linked the cities of its name and ended at President Street in Baltimore, on the east side of the "Basin"; there it joined B & O's line across Pratt Street (in the earlier days, PW & B passenger trains used the B & O Charles Street station).

By 1840 the three railroads covering the New York-Philadelphia portion of the route were jointly operating through passenger trains. In fact two of them, the Camden & Amboy and the Philadelphia & Trenton, already were part of the infamous "Joint Companies," the combination of commonly-controlled railroad and canal companies which monopolized all transportation between New York and Philadelphia. The "Joint Companies" went on to absorb the New Jersey Rail Road & Transportation Company in 1867, putting all carriers on the route into one family. Thus, except for the ferry, travelers between New York and Philadelphia usually could ride straight through. But from there south were several irritating hurdles.

Philadelphia itself was the worst. Whichever way the traveler arrived from New York, he had a long way to get to the PW & B station. That two and one half-mile gap between the Philadelphia & Trenton's Kensington station and the PW & B terminal at 11th & Market Streets was bad enough, but in 1852 the PW & B proceeded to widen it to three and one half miles when it built an imposing new station to the south at Broad and Prime Streets (Prime Street is now Washington Avenue). Then, between Wilmington and Baltimore, PW & B trains had to be rolled onto a car-ferry to cross the wide Susquehanna River at Perryville and Havre de Grace, Maryland. Finally, in Baltimore the trains were broken up in the PW & B's President Street yard and each car was individually hauled by horses a mile along Pratt Street to the B & O station at Charles Street. And when it came time to leave Baltimore for Washington, horses again took each car another mile west on Pratt Street to Mt. Clare, where they were assembled into a train. The animal power was necessitated by city ordinances prohibiting steam locomotives in downtown streets.

Nor were train schedules completely coordinated. In 1851, for example, one could leave New York on the 6 a.m. ferry and get to Philadelphia (Kensington, that is) at 11:15 — plenty of time to make the PW & B's 2 p.m. departure for Baltimore. (A gambler could sleep later and leave New York at 9 a.m., but would have only fifty minutes to get across town in Philadelphia — if his train was on time.) The 2 p.m. PW & B train arrived in Baltimore at 6:30, but there the traveler had to give up for the night, since the last

PASSENGERS and FREIGHT

TO THE

And SOUTHWEST!

FROM NEW YORK

TO

**CINCINNATI, INDIANAPOLIS, LOUISVILLE,
ST. LOUIS, MEMPHIS, WHEELING,
COLUMBUS, DAYTON, TOLEDO, CHICAGO,**

BY THE

**BALTIMORE & OHIO
RAILROAD!**

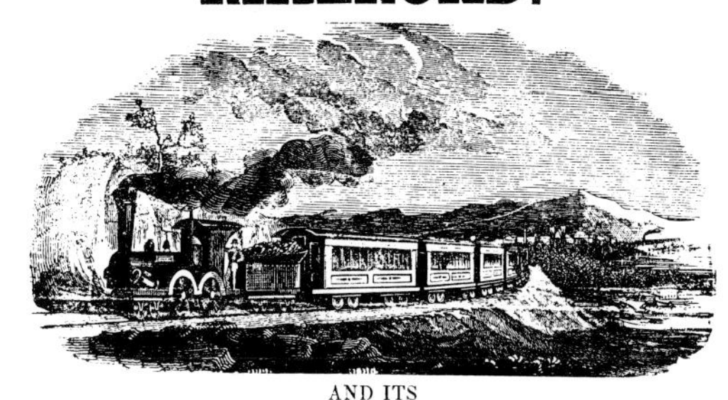

AND ITS

EASTERN AND WESTERN CONNECTIONS!

Travelers are informed that full arrangements have now been effected to ticket them through from New York to the West, by direct Railroad connection, and yet affording a visit to Philadelphia, Baltimore, Harper's Ferry, Cumberland, Wheeling, etc., etc , on the way.

Tickets bought in New York for the West, by this route, will enable the traveler to lie over at the above named places, and also afford time, at Baltimore, to visit Washington City, but 40 miles distant.

THE PRINCIPAL CONNECTING TRAINS for this route leave foot of Courtland Street, New York, at 8 and 10 A. M. and 6 P. M. daily, except Sundays, when the 6 P. M. train leaves only. The 8 A. M. train connects at Baltimore with the 5 P. M. Express train of the Baltimore and Ohio Railroad for Cincinnati, etc., direct.

The cost of through tickets by this route will be the same as by the great Railroad routes of New York to the West.

The most desirable connections are formed by this route, at Louisville, with the first and second class Steamers for the Mississippi and all other Southwestern Rivers.

THROUGH TICKETS to be had at the Office of the New Jersey Railroad Company, foot of Cou-tland Street, New York.

FARE:

**From New York to Wheeling, $14.00; to Cincinnati, $18.50
" " " Columbus, 16.40; to Indianapolis, 20.00
" " " Louisville, 21.00; to Dayton, 18.00
" " " " (by Steamer from Cincinnati,) 20.50**

For further particulars, apply to

J. L. SLEMMER, Agent B. & O. R. R.
Corner of Broadway and Park Place.

FREIGHTS TO THE WEST.

This road (and its connections) offers a very desirable route for the shipment of all freights between Boston and New York and the entire West and Southwest. Through receipts to Wheeling at the lowest possible rates, and guaranteeing the transportation in the shortest possible time, will be given by H. B. CROMWELL & CO., Agents, at Office of Baltimore Steamship Line, corner of Albany and Washington Streets, New York, and by J. W. POTTLE & CO., Agents of Merchants and Miners' Line, Central Wharf, Boston.

At Wheeling (on the Ohio River 94 miles below Pittsburg) the amplest facilities are afforded for the re-shipment of freights to all points farther west, and at rates as low or lower than from Pittsburg.

J. H. DONE, Master of Transportation,
Baltimore and Ohio Railroad, Baltimore.

"The Printing Office," Steam Press, Sun Iron Building, Baltimore.

This handbill, dating to about 1855, advertises through ticketing between New York and western points. It was hardly a through train service, however; at least two transfers were necessary to reach Wheeling, and more for the other points listed. H. H. Harwood, Jr. Collection.

B & O's line to Washington looked like this in the late 1850s. The scene is the Washington branch junction at Relay, Maryland in 1857. Smithsonian Institution Collection.

B & O Washington train left at 5 p.m. Going north, however, an early riser could accomplish the entire trip in a day by leaving Washington at 6 a.m. and finally staggering off the ferry at Cortlandt Street in New York at 9 p.m. But along the way he at least had a leisurely two and one half hours to change stations at Philadelphia.

Slowly and erratically, services became better coordinated and facilities improved. Better stations seemed to come first, before better connections. In 1850 the PW & B moved out of the B & O's makeshift Baltimore station at Pratt and Charles Streets and into a new Classical-style terminal of its own on President Street, although the horse-drawn transfer between the two railroads continued essentially as it had before. Two years later the PW & B opened a larger and more ornate version of its Baltimore station in Philadelphia at Broad and Prime Streets, eliminating a tedious stretch of horse-drawn car hauling to its old (but better located) Market Street terminal. In Washington the B & O finally replaced its one-time Pennsylvania Avenue boardinghouse with a handsome Italianate terminal three blocks north at New Jersey Avenue and C Street. Opened in 1851 and completed in 1852, the new Washington station also eliminated a short stretch of horse-powered hauling.

And at Baltimore the B & O realigned its passenger route into the city in 1853, bypassing a mile of the horse-drawn street operation there; in 1856 it completed the first section of its ostentatious Camden station at the end of the new line. The horses were not completely gone yet, however; they still took the cars one by one for the mile across Pratt Street between Camden and the PW & B's President Street station. By 1859 the B & O and PW & B were operating some through Philadelphia-Washington services, including sleeping cars rebuilt from PW & B coaches.

The Civil War and the dawning of railroad power politics helped accelerate the creation of connecting trackage and the start of through New York-Washington train services. The war instantly turned Washington into a true capital — and a militarily vulnerable one at that. It was now an urgent destination for troops, politicians, civilians of all sorts, supplies, and high-priority mail. The disconnected and disjointed state of the route ceased being merely irritating and became seriously threatening. A "normal" trip between New York and Washington took between eleven and twelve hours and, since most of the route was single track, the flood of new traffic often created delays.

And, if nothing else, the country got a brutally graphic illustration of the route's shortcomings with the infamous "Pratt Street Riot" in Baltimore on April 19, 1861 which ended with the Civil War's first fatalities. (And which was, ironically, a clash between soldiers and civilians.) A regiment of 700 soldiers being moved from Massachusetts to Washington was brought into Baltimore by the PW & B; there, of course, their 35 coaches had to be hitched to horses and dragged the mile over Pratt Street to the B & O's Camden station. The transfer immediately turned into a shambles as a large, hostile crowd of Confederate sympathizers tried to block the cars. It then turned bloody as the remaining troops, forced to march in the street, began shooting their way through the mob. Twelve civilians and four soldiers died, and many were wounded.

Within months after the war began, pressures built for government operation of the New York-Washington route; others, sensing opportunity, began proposing new lines. In January 1862 Congress passed an act authorizing the President to take over the various railroads if and when needed. Fast behind this came a proposal that the government build a separate new double-track "Air Line" railroad from New York to Washington. ("Air line" in those days meant merely a straight, or mostly straight, route.) In late 1862 a formal "Air Line" bill appeared in Congress. Although it was a popular idea, the real hand behind the bill was the powerful Pennsylvania politician Simon Cameron who also happened to be an open ally of the Pennsylvania Railroad. Although not directly a part of the New York-Washington route, the Pennsy already was developing designs on some of the markets.

The B & O's John Garrett controlled only the short Baltimore-Washington segment of the New York route, but

One of the first truly elaborate "permanent" stations on the New York-Washington route was the PW & B's Baltimore terminal at President Street, opened in early 1850. Located at the eastern edge of the Inner Harbor, it amazingly survived into 1990, albeit as a damaged and stripped shell. Smithsonian Institution Collection.

he quickly saw that he had the most to lose from any government takeover or competitive line. When the Washington branch was built, the State of Maryland had guaranteed the B & O a rail monopoly between Baltimore and Washington — which also gave Garrett a strategic advantage in reaching farther south once the war ended. He thus took the lead in lobbying against the new "Air Line" and in persuading his partners on the New York route to improve service. At about the same time, the four major railroads entering Philadelphia, the PW & B, the Pennsylvania, the Philadelphia & Reading, and the Philadelphia & Trenton, realized the commercial advantages in connecting themselves better.

These converging pressures produced results, although not immediately. First to appear was an attempt at the first genuine regularly-scheduled through passenger operation between New York and Washington — necessarily an awkward one, since there was still no suitable connecting rail route through Philadelphia. In May 1862 a short-lived service was started using the Camden & Amboy line between Trenton and Camden; at Camden the cars were floated across the Delaware on a barge and delivered to the PW & B's track on the Philadelphia waterfront at the foot of Washington Avenue. Freight was also ferried over this route.

In the meantime work had started on an all-rail connection through Philadelphia to link the four main line railroads and, as one purpose, to allow direct train transfers between the Philadelphia & Trenton and the PW & B. A company called the Junction Railroad had been formed to do this back in May 1860, but it was not until 1862 that it settled on a final route. Not an operating company itself, the Junction Railroad was an early form of joint facility which was equally owned by three of Philadelphia's main line railroads: the PW & B, the Pennsylvania, and the Philadelphia & Reading (more commonly called the "Reading"). Each of the three lines ran its own trains over the Junction's tracks. Its three-mile-long line began on the

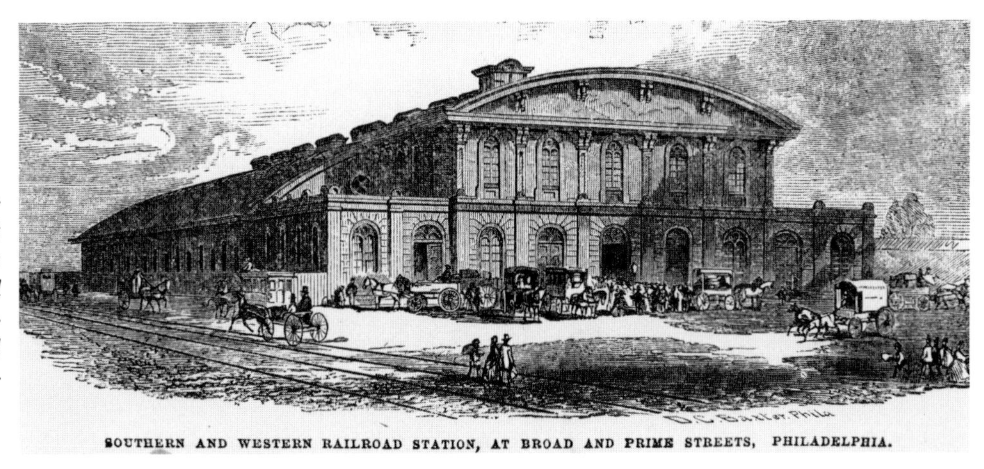

In 1852 the PW & B relocated its Philadelphia terminal to this ornate new structure at Broad and Prime Streets, on what was then the city's far south side. Although abandoned for passenger use by the early 1880s, it survived in altered form as a freight station until the mid-20th Century. Smithsonian Institution Collection.

B & O's bid for status in its home city was Camden Station, first opened in 1856. This photograph shows it as fully completed after the Civil War. The central section is the 1856 building; the two wings were added in 1865. In the following years it was subjected to numerous alterations, but still stands. B & O Museum Collection.

north at Belmont in Fairmount Park where it met the Reading; it then followed the west bank of the Schuylkill River south to a connection with the PW & B at Gray's Ferry in the southwest part of the city. Midway it met the Pennsylvania near 35th Street and Mantua Avenue, and used the Pennsy's tracks for about a mile through West Philadelphia. And there lay a future joker.

The Junction project went slowly, but by November 1863 enough connecting track had been finished so that there was a stopgap all-rail passenger service through Philadelphia. Trains from New York (actually Jersey City, of course) were switched from the Philadelphia & Trenton to the Reading's Port Richmond branch in Kensington. From there they were hauled west across the northern part of town and over the Schuylkill to West Falls, then reversed and taken south to Belmont. Joining the Junction Railroad at Belmont, they continued down the west bank of the Schuylkill to West Philadelphia, then were switched onto the Pennsylvania's newly-built Delaware Extension and once again crossed the river, now heading eastward. At

B & O's fashionable and aesthetic Washington terminal opened onto unpaved streets when it was completed in 1852. Located at New Jersey Avenue and C Street, it was within two blocks of the Capitol. Altered and enlarged in 1889, the building served all B & O Washington passengers until Union Station opened in 1907. B & O Museum Collection.

25th and Washington in South Philadelphia they were rolled onto PW & B rails and taken a short distance east to the PW & B's Broad Street terminal.

By December 1864 another portion of the Junction Railroad was finished and the Washington trains could be run directly down the Schuylkill line to the PW & B at Gray's Ferry, avoiding the eastward jog over the Pennsy's Delaware Extension. In 1865 four through passenger trains each way were advertised between New York and Washington, using this seesaw path through Philadelphia; the fastest took nine hours and fifty minutes. (All elapsed times traditionally were based on the ferry arrivals and departures at New York City rather than train times at Jersey City. The difference generally amounted to ten to fifteen minutes.)

The Philadelphia connecting trackage project finally was completed in 1867 with the opening of a new line directly linking the Pennsylvania and the Junction Railroad with the Philadelphia & Trenton — and at last eliminating the backward-forward haul over the Reading's Port Richmond branch. Called the Connecting Railway (and which had no corporate tie to the Junction Railroad), it started near the present "Zoo" junction, crossed the Schuylkill, and ran northeastward across the city's north side to join the Philadelphia & Trenton at Frankford Junction. Using this route (which is currently Amtrak's main line through North Philadelphia), trains now had a straight movement through Philadelphia between the Philadelphia & Trenton and the PW & B. But significantly, the Connecting Railway line to Frankford Junction was not a three-way joint project like

More typical of the primitive early passenger facilities along the New York rail route was the Camden & Amboy's Trenton, New Jersey station, shown here about 1865. W. R. Osborne Collection.

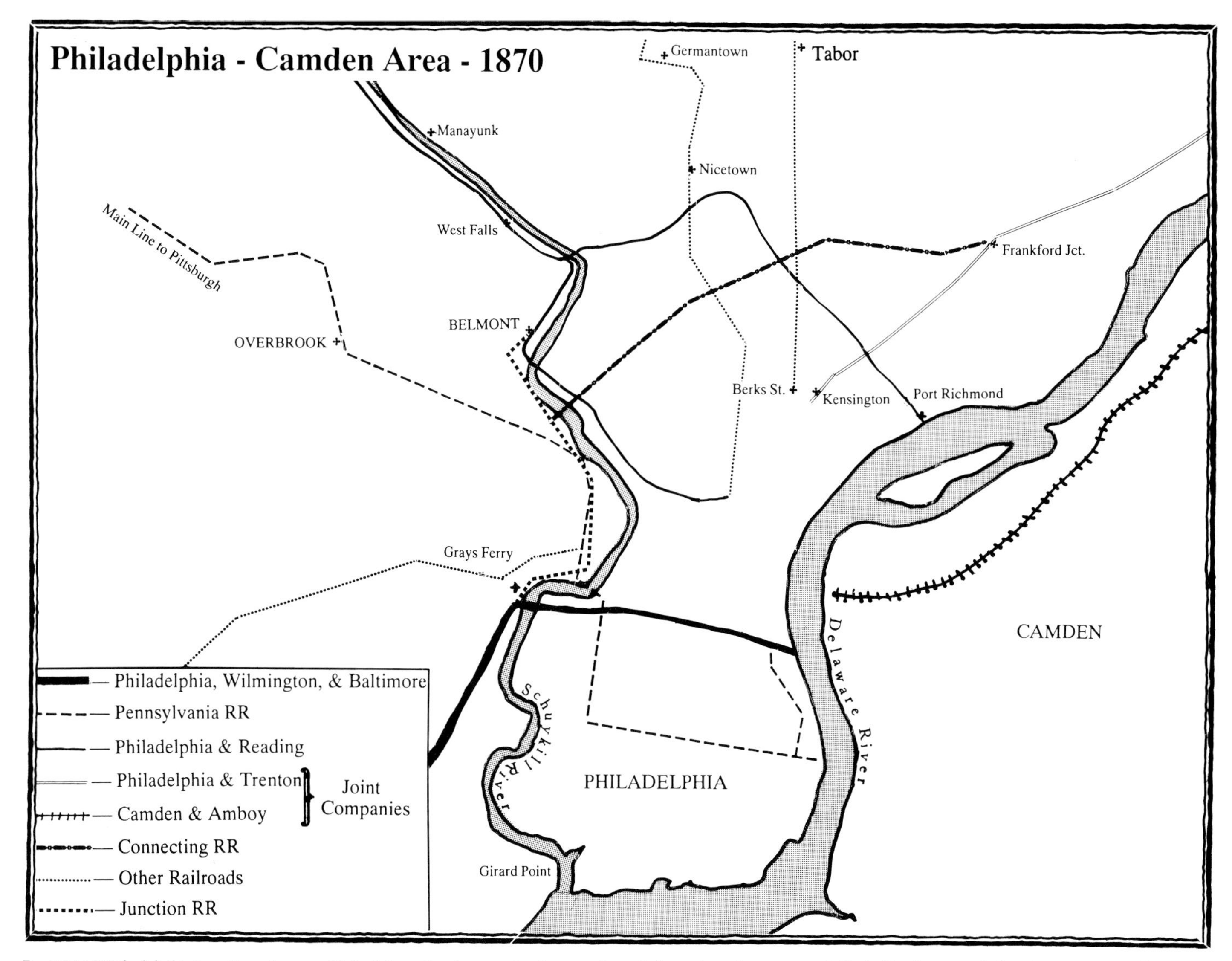

By 1870 Philadelphia's railroads were linked together by a pair of separate rail lines forming a sort of "belt line" around the city. The jointly-owned Junction Railroad runs roughly down the left side of this map; the PRR-financed Connecting Railway extends across the top as far as Frankford Junction.

the Junction; it had been entirely financed by the Pennsylvania and leased to the Philadelphia & Trenton.

As the Philadelphia problems were painfully being solved, the PW & B eliminated its own major bottleneck. After one false start and four years of effort, it finally conquered the Susquehanna River in 1866 with a magnificent 3,200-foot thirteen-span wooden truss bridge. And elsewhere along the New York-Washington route the various railroads finished the job of double tracking, making a few realignments in the process. Between 1862 and 1864 the Camden & Amboy phased in an almost completely relocated double-track route between Trenton and New Brunswick, New Jersey, and the Philadelphia & Trenton was double tracked in 1865. The B & O had finished double tracking its Washington branch in 1864.

Thanks to all these improvements, transit times quickened, but were still leisurely by anyone's standards. In 1868, with all the Philadelphia connections opened and the Susquehanna bridged at Havre de Grace, through passenger cars could make the trip in nine hours and ten minutes. Two years later one could leave Manhattan on the 8:40 a.m. ferry from Cortlandt Street and be delivered to the B & O's New Jersey Avenue terminal in Washington at 5:20 p.m., eight hours and forty minutes later.

Through all this development, the railroads making up the New York-Washington route remained independent working partners, each contributing cars and splitting the revenues among them. B & O was merely one of the partners, with no financial interests in the other railroads or direct management control over them. It neither needed to nor wanted to. If the through service was still somewhat sluggish, the multi-railroad arrangement seemed stable and secure. Except for the B & O, the various separate railroads were relatively short regional companies without conflicting interests elsewhere. None of them competed with one another and each had a monopoly in its own territory. (Garrett and his partners had managed to beat down the federal "Air Line" proposal in 1864 and killed off a revived version in 1867.) So it was obviously in everyone's best interest to work together.

Thus the B & O had the best of all worlds. It had an outlet to Philadelphia and New York, and at least some control over the services and rates, but with no heavy financial commitments. John Garrett could devote most of his attention and the B & O's somewhat limited financial resources to the railroad's real mission — building Baltimore's western and southern trade. Or so he thought.

By the early 1870s, however, the wider railroad world was rapidly changing and turning turbulent. Gradually, large competing territorial systems were being welded together, and new rail lines were being built into what were once the private preserves of the older railroad companies.

III

NEW YORK
AND
Washington Air Line Railway,

Between N.Y. City & Washington, D.C.

Via Trenton, Philadelphia, Wilmington & Baltimore.

This Line now has a double track throughout, and embraces Roads of the following Companies, viz:

NEW JERSEY R.R., A. L. Dennis, Pres., N.Y. City,
CAMDEN & AMB'Y R.R., W.H. Gatzmer, Pres., Phil.
PHIL. & TRENT'N R.R., V. L. Bradford, Pres., Phil.

Ashbel Welch, Gen'l President, Lambertville, N. J.
F. Wolcott Jackson, Gen. Sup. Jersey City.

PHILADELPHIA, WILMINGTON, AND BALTIMORE RAILWAY,
Isaac Hinckley, Pres., Philadelphia; H. F. Kenney, Sup., Philadelphia.

BALTIMORE AND OHIO, (Washington Branch),
John W. Garrett, Pres., Baltimore; John L. Wilson, Mas. Trans., Baltimore.

ACTING TERMINAL AGENTS:

G. W. BARKER, (Master of Transportation, New Jersey R. R.), - - - - - - - - - - Jersey City.
GEORGE S. KOONTZ, (Agent Baltimore and Ohio R. R.), - - - - - - - - - - Washington.

WINTER SCHEDULE, TO TAKE EFFECT DEC. 30, 1869.

NEW YORK TO BALTIMORE AND WASHINGTON. PHILADELPHIA TO WASHINGTON.						WASHINGTON TO PHILADELPHIA AND NEW YORK. BALTIMORE TO NEW YORK.				
Morning Express.	Noon Acc.	Phila. & Wash. Express.	Night Express.	Miles	STATIONS.	Fares	Morning Express.	Noon Acc.	Phila. & Wash. Express.	Night Express.
A. M.	NOON.		P. M.		Lve.......... **New York** .... Arr		P. M.	P. M.		A. M.
8 40	12 30		9 20		Cortlandt Street Ferry		4 44	10 20		6 09
8 52	12 40		9 32	1	Jersey City		4 29	10 10		5 59
9 16	1 05		9 56	9	Newark		4 04	9 45		5 34
10 04	1 54		10 46	32	New Brunswick		3 19	8 54	P. M.	4 38
10 49	2 44	P. M.	11 33	58	Trenton		2 32	8 01		3 49
12 04	4 09	11 30	12 54	90	Philadelphia		1 14	6 39	11 24	2 29
		Broad St.			*W. Market Street*				Broad St.	
1 09	5 21	12 52	2 07	119	Wilmington		12 14	5 31	10 14	1 26
2 17	6 38	2 05	3 16	152	Perryville		11 00	4 04	8 56	12 08
3 44	8 13	3 50	4 49	189	Baltimore		9 34	2 34	7 24	10 39
					President Street					
5 20	10 10	5 10	6 35	228	**Washington**		8 00	12 45	5 40	9 00
P. M.	P. M.	A. M.	A. M.		ARRIVE.] [LEAVE.		A. M.	NOON.	P. M.	P. M.

Sunday Trains.—The Night Express between New York and Washington and New York and Baltimore, and the Philadelphia and Washington Express will run daily, including Sundays.

[The time given in the foregoing tables is *New York time* for Southward, and *Baltimore time* for Northward trains, there being ten minutes difference in longitude.]

The Morning and Night Express Trains are exclusively for Through Travel, excepting that the Southward trains take passengers from New York for Wilmington, Perryville and Baltimore; and the Northward Trains from Washington and Baltimore, for Wilmington, Trenton, New Brunswick, Newark and Jersey City. The Morning Express also takes passengers between Broad Street Station, Philadelphia, and Washington, each way.

The Noon Trains in each direction are the Through Accommodation, and take passengers for all important way points between New York and Washington. (See local advertisements.)

All the trains stop ten minutes at Wilmington for lunch—the day train porters taking orders from ladies for luncheon, to be brought into the cars if required.

Through tickets, baggage checks, sleeping car berths, seats in the reclining chair cars, and reserved seats or saloons in day compartment cars, may be had at Dodd's Express office, 944 Broadway near 23d street, and at Cortlandt Ferry Ticket Office, New York; No. 828 Chestnut street, in Continental Hotel, Philadelphia; at No. 147 Baltimore street, Baltimore; and at 406 Pennsylvania avenue, Washington.

Through Conductors, or Baggage Agents, will accompanying the through trains, to afford every attention to passengers, in providing seats, information as to connecting lines, city cars, and other conveyances, care of baggage, etc. Through colored porters will attend to the comfort of passengers, keep the cars cleanly, and to do other offices that may be appropriately required of them, and without fee.

In order to secure Prompt Delivery of Baggage in New York arriving by the Washington trains, Dodd's Express Agents will accompany the night trains through from Washington, and collect checks before the passengers retire for sleep.

Any difficulty experienced by travelers in using the line, and especially *any neglect or want of care or civility* on the part of any person connected with its operation, should be at once made known by letter, with date, train and other particulars, addressed to the Gen'l Manager.

W. P. SMITH, Gen'l Manager, Washington.

The New York-Washington route was not much of an "air line" in 1869, but was much improved over what it was seven years earlier. Although this advertisement carefully creates the illusion of a single railroad, it actually was a joint service of the five companies shown. (The New Jersey Railroad, Camden & Amboy, and Philadelphia & Trenton were under a common control, however.) The route's general manager, William Prescott Smith, had been B & O's master of transportation and returned to that job in 1872.

Typical of 1860s-era passenger equipment on the New York-Washington route, a Philadelphia-bound New Jersey Railroad train poses outside Jersey City in 1867. The Rogers-built 4-4-0 heads six clerestory-equipped cars. C. B. Chaney Collection, Smithsonian Institution.

And once started, the process fed on itself as all railroads scrambled to protect their old markets and tie down new ones before some rival did. The most aggressive was the Pennsylvania Railroad under J. Edgar Thomson and his executive vice president, Thomas A. Scott. By 1873 the PRR had transformed itself from a single line to Pittsburgh into a trunk system covering much of the East and Midwest. From its base at Philadelphia it reached Chicago, St. Louis, Cincinnati, Louisville, four Great Lakes ports, and portions of Michigan and upstate New York. And, in one of his earlier moves, Thomson and his political ally, Simon Cameron, had bought working control of the Northern Central Railway in 1861 — giving the Pennsy a back-door entrance to the B & O's bastion at Baltimore. Not far behind Thomson and Scott were Cornelius Vanderbilt and his son William, fast putting together what became the New York Central system covering much the same territory.

By then, too, John Garrett had an absolute hold on the B & O. Having guided the railroad through its many wartime traumas and rebuilding it afterwards, Garrett became the architect of its postwar expansion — the B & O's equivalent of Thomson, Scott, and the Vanderbilts.

But Garrett was a different personality with different strategic goals. Although the B & O grew considerably under him, it did so more fitfully, slowly, and defensively than the Pennsy and Central. Viewed from the hindsight of history, he seems to have had more caution and less vision than his peers. Along with this came a method of financial management which promised future problems. In short, Garrett seemed dedicated to preserving Baltimore's (and his own) financial control of the railroad. As a result he resisted issuing new stock to help finance his extensions and acquisitions, since doing so would dilute control and perhaps allow "outsiders" to influence the B & O. Instead he borrowed money and issued bonds, slowly building up a dangerously large debt. In addition he paid dividends often when they were not justified. The end result for the B & O was a constant shortage of ade-

Fresh from the Hayward & Bartlett plant in Baltimore, the PW & B's No. 51 Phantom poses for its official picture in front of B & O's adjacent Mt. Clare station in 1865. Phantom's ornate trim includes a raised eagle medallion between its drivers. C. B. Chaney Collection, Smithsonian Institution.

quate capital which produced an oddly uneven type of development. Line expansions frequently seemed to be financed at the expense of rebuilding and modernizing the railroad's physical plant, and the expansion itself sometimes went in peculiar directions.

By 1874 Garrett's B & O had begun to look quite impressive — on a map, at least. Since 1857 the railroad had a working route to Cincinnati and St. Louis, although it did not yet fully control the western sections of the line. In 1871 Garrett had managed to invade the Pennsylvania's stronghold at Pittsburgh, and in 1874 he opened a line into Chicago. He also was attempting to build southward up the Shenandoah Valley toward Roanoke, Virginia and had completed the Metropolitan branch into Washington from the west.

So, in theory, the B & O was a contender in the same general territory as the Pennsy and the growing Vanderbilt system. But in reality it was a weaker contender, with more difficult and sometimes roundabout routes, poorer market penetration, and shakier finances. It was, and always would be, third among the "big three" eastern rail systems. To compensate and attract business, Garrett periodically resorted to rate cutting, which especially aroused the Pennsy's enmity. Between their natural commercial rivalries and Garrett's maverick maneuverings, the PRR and B & O developed an enthusiastically hostile relationship.

Nowhere were these power struggles better reflected than along the New York-Washington route, nor was there a better showcase of the differing abilities, strategies, and styles of the leaders of the two railroad systems. Soon after the end of the Civil War the Pennsylvania began moving. The events which followed were complex, often jumbled, and occasionally farcical, but the end result was all too clear: beginning in 1871 with no real operating presence along the New York-Washington corridor, the PRR emerged ten years later not only spanning the entire route, but owning the choicest lines. By 1881 Garrett had been left isolated with only his Baltimore-Washington branch — and in a deep strategic quandary.

The first seeds were sown in 1867, about the time that Garrett and his cohorts had managed to defeat the PRR-supported "National Air Line" project for the final time. Undeterred, the Pennsy's Thomson and Scott merely changed tactics and scored an especially spectacular coup on Garrett's southern flank. By picking up the charter of the unbuilt Baltimore & Potomac Railroad and using an unnoticed and seemingly innocuous clause in it, they were able to begin building a Baltimore-Washington line of their own which, to Garrett's dismay, legally bypassed the B & O's state-protected monopoly of the Washington business. Completion of the new Pennsy line would take six more years, but building was well under way when the PRR moved again — this time at the north end of the route.

John W. Garrett, as he looked in his commanding and portly prime. Clearly he was not to be questioned, particularly by subordinates. B & O Museum Collection.

By then the perceptive Pennsy managers had recognized what Garrett did not see soon enough — that New York was the key to power in the railroad world. Its 1870 metropolitan area population was 1.5 million, by far the country's largest, and it was handling 57 percent of the nation's foreign trade. Admittedly, Thomson and Scott had to shed their own parochial blinders, which they did not do instantly. But since the early 1860s the Pennsy found itself increasingly involved in the New York market, initially pushed into it by its customers and by its western connections. Like the B & O, the PRR was routing its New York business over the "Joint Companies" east of Philadelphia; but unlike the B & O, it came to realize the necessity of having full control of its New York entryway. In June of 1871 it signed a lease of the "Joint Companies" and directly assumed operation of the three railroads linking Philadelphia and New York. In the bargain it also got the only other major carrier on the route, the Delaware & Raritan Canal. The Pennsy's primary purpose was to tie down a New York-New Jersey terminal for its main line western traffic, but at the same time it was now established at both ends of the New York-Washington corridor. It also now controlled Garrett's New York entrance.

When it arrived in Washington in 1872, the Pennsylvania built this ostentatious passenger terminal on the Mall at 6th and B Streets, now the site of the National Gallery of Art. L. W. Rice Collection.

On the south end, the PRR finally completed its Baltimore-Washington line in 1873; by that time, Garrett found himself in even deeper trouble. In the process of building its line, the Pennsy had taken care to undercut the B & O's working connections at both Baltimore and Washington. In Baltimore the project included an extension eastward through the city to join the still-independent Philadelphia, Wilmington & Baltimore at Bay View, on Baltimore's far east side. This connection avoided the plodding horse-drawn car transfer necessary for all B & O-PW & B traffic; it also bypassed two miles of slow in-street running over the inner part of the PW & B's own main line into Baltimore on Boston and Fleet Streets. Using the Bay View connection and its new Washington line, the Pennsy now could operate its own through New York-Washington trains via the PW & B — and could move them through Baltimore far faster than the competitive B & O-PW & B route.

Garrett's woes at the Washington end were worse. Besides being a growing market in its own right, although mostly a passenger market, Washington was even more important as the only major eastern gateway to the South. Both Garrett and the Pennsy's Tom Scott had ambitions to extend southward and were maneuvering to control some of the now-prostrate railroads of the old Confederacy. But again it turned out to be no contest. In a deft, albeit shady political coup (engineered by Senator Simon Cameron), the Pennsylvania obtained exclusive rights to use the government-built Long Bridge over the Potomac at Washington, the sole rail crossing into Virginia. It then proceeded to establish direct connections with the two rail lines reaching south: the Virginia Midland (now Southern Railway) at Alexandria, Virginia and the Richmond, Fredericksburg & Potomac at Quantico. The B & O, which ironically had just bought control of the Virginia Midland, was effectively cut off from all southern connections.

Garrett tried to recover by setting up an awkward, expensive, and only partly effective freight carfloat across the Potomac to his Virginia Midland at Alexandria. But the RF & P was completely beyond reach, as was any practical hope of operating a North-South passenger service over either line. It was not until 34 years later that B & O freight trains were allowed to cross the Long Bridge to the newly-built Potomac Yard and have direct rail access to the southern network. And never again would B & O be able to participate in the ever-growing through passenger routes between the South and the northeastern seaboard.

Despite the Pennsy's decisive new presence in the territory, B & O's New York-Washington business chugged on as usual. The PRR made no immediate move to disrupt the old agreements between B & O and the "Joint Companies" between Philadelphia and New York, and the PW & B remained independent and friendly to everybody. But at the same time the Pennsy, also using the PW & B, was now a direct competitor over the entire route. Garrett clearly stood in the eye of a hurricane; deceptively peaceful, it was also likely that something else would be blowing his way soon.

Although nothing quite so violent happened, inevitably there came a series of harassments. Reacting to the B & O's entry into Chicago in 1874 and its subsequent rate cutting, the Pennsy periodically cut off the Philadelphia-New York end of Garrett's through passenger service. Court orders or shaky peace treaties restored the B & O operation, but Garrett obviously had to find another way to get to New York. In the mid-1870s, however, there was none.

Perhaps not by coincidence, an alternative route quickly began to materialize. In 1874 the North Pennsylvania Railroad, an independent company operating from Philadelphia to Bethlehem, Pennsylvania, began building a New York branch. The new North Penn line left its main line at Jenkintown, a suburb ten miles north of Philadelphia, and swung northeast to the Delaware River at Yardley, Pennsylvania. From the Delaware River an affiliated company, the Delaware & Bound Brook, continued the route across New Jersey as far as Bound Brook, New Jersey, thirty miles short of the Hudson River at Jersey City. At Bound Brook it joined the main line of the Central Railroad of New Jersey, another independent company which crossed the state westward from Jersey City to Easton, Allentown, and Wilkes-Barre, Pennsylvania. The CNJ's already-existing double-track line completed the new route to Jersey City, where its ferries shuttled across the Hudson to the foot of Liberty Street in lower Manhattan.

Opened May 1, 1876, in time for the Philadelphia Centennial Exposition, the joint North Penn-Delaware & Bound Brook-Jersey Central route was a well-built high-speed line which broke the Pennsylvania's Philadelphia-New York monopoly. It was less competitive at intermediate points, however, since it bypassed Newark, New Jersey's largest city, and the state capital at Trenton — both of which were on the Pennsy main line. Passengers and freight on the new route could reach both Newark and Trenton over short branches, Trenton from Trenton Junction (later called West Trenton) and Newark via CNJ branches out of Jersey City and Elizabethport, New Jersey; but the services were never truly competitive.

Advertised as the "Bound Brook Route" in honor of the Jersey Central junction point, the new Philadelphia-New York service operated independently for three years after it was opened; in the meantime B & O's New York business continued moving over the PRR east of Philadelphia. But in 1879 the cauldron began bubbling again. In May of that year the aggressively expanding Philadelphia & Reading leased both the North Pennsylvania Railroad and the Delaware & Bound Brook. (The Jersey Central remained independent, but a close working partner. Later the Reading made two short-lived attempts to lease it and finally bought stock control in 1901.) In the meantime, Garrett had quietly linked up with the Reading's ambitious president, Franklin Gowen, and was negotiating to use the Bound Brook Route to New York.

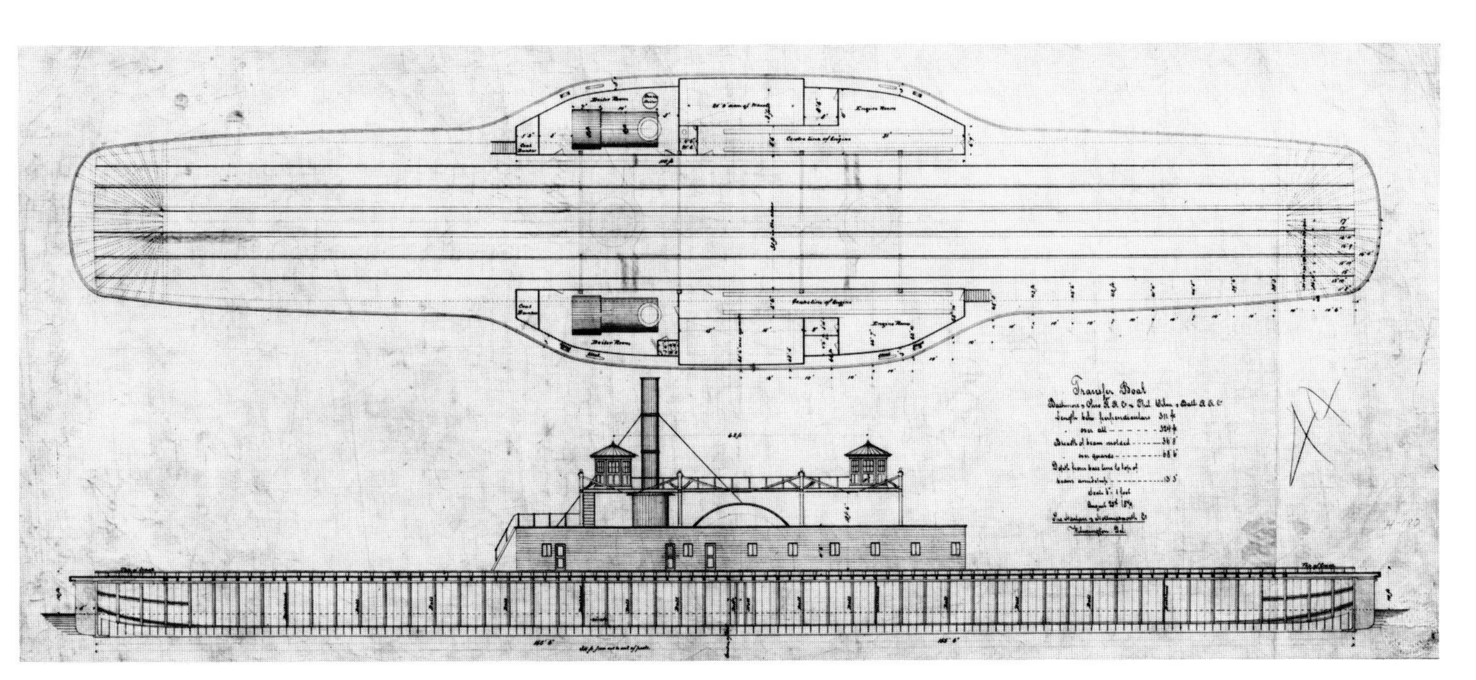

In 1879 the B & O and PW & B jointly ordered the trainferry Canton from Wilmington shipbuilder Harlan & Hollingsworth. It shuttled cars across Baltimore Harbor from 1880 to 1886, then was briefly used by the PRR in New York Harbor. Like many such vessels, it ended life as a barge. Hagley Library Collection.

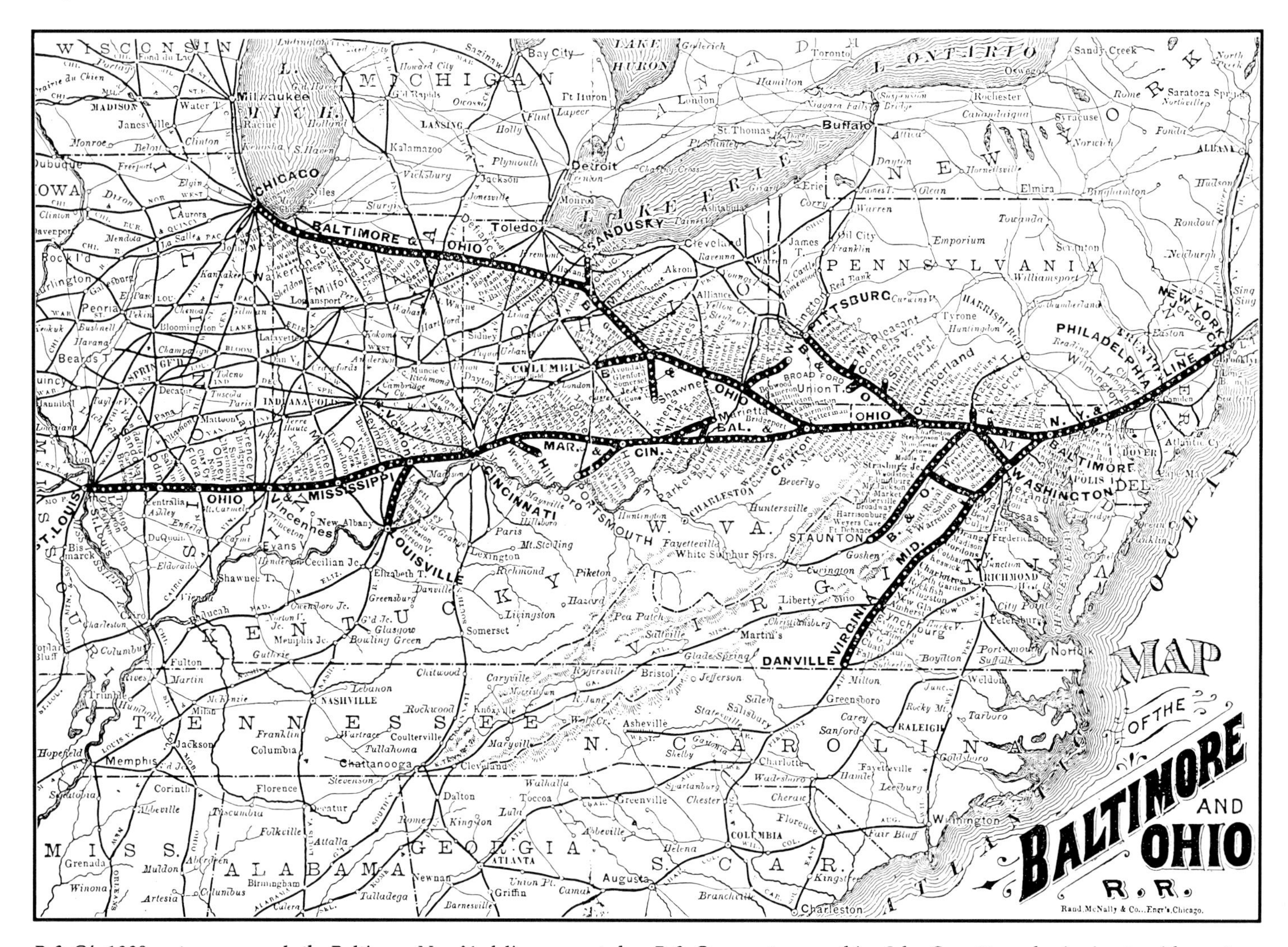

B & O's 1880 system map made the Baltimore-New York line appear to be a B & O property, something John Garrett was beginning to wish was true. The route shown included the "Joint Companies" between Philadelphia and New York, although when this timetable was issued on December 19, 1880, B & O had switched to the Bound Brook route. Note also the "B & O" line to Danville, Virginia, which actually was owned by the B & O-controlled Virginia Midland (now the Southern Railway). B & O Museum Collection.

But he made no overt moves yet. Both physical and legal work were needed before the B & O could effectively use the line. The old North Penn had to be more directly linked to the Reading in the outskirts of Philadelphia; this was accomplished in early October 1879 when a two-mile connecting line was opened between Wayne Junction on the Reading and the North Penn at Tabor.

A more nagging legal problem had to be resolved. B & O traffic was still carried by the PW & B between Baltimore and Philadelphia. But to get from the PW & B to either the PRR or the Reading in Philadelphia, its cars had to move over the Junction Railroad — the jointly-owned connecting line which had been built back between 1863 and 1866.

When the Junction Railroad was built, the PRR's main line already occupied about a mile of its planned route through the West Philadelphia area. Rather than build a separate line, the three Junction Railroad partners agreed to use the Pennsy trackage as part of the joint facility. But when the last piece of the Junction project was finished in 1866, the Pennsy surprised its other two partners by claiming full ownership of this section and, after a lawsuit, it was later upheld in court. Although the Junction continued operating as a commonly-used connecting line, the Pennsylvania had the power to cause trouble if it ever wanted to. And it decided to when some New York freight began moving between the PW & B and the Bound Brook Route.

To settle the question, another lawsuit was necessary. In April 1879, even before the Reading's lease of the North Penn was announced, the PRR and the Junction Railroad were jointly sued by individuals representing the Jersey Central and the Delaware & Bound Brook, with B & O's open support. (As part-owners of the Junction Railroad, the Reading and PW & B stayed out of the suit, although the Reading's Franklin Gowen was clearly behind it.) Their purpose was to prevent the Pennsy from interfering with any interchange traffic going to or from the Bound Brook Route — which, of course, included any B & O business which might use the line.

The suit took over a year to be decided. But finally, on October 28, 1880, the court ruled that the Pennsy had to operate its short connecting section in good faith for all traffic. It left a loophole, however, by allowing the railroad to use its own locomotives to move anything over the controversial mile of track.

Once it received a legal green light, the B & O moved quickly. Ignoring the ninety-day cancellation clause in its Philadelphia-New York agreement with the PRR, it abruptly switched its passenger trains to the Reading-operated Bound Brook Route December 1, 1880. Freight followed a month later.

Although probably not surprised, the Pennsy did its best to make the changeover as miserable as possible for the B & O. It insisted on its engine change for its short section of the Junction Railroad, it collected a small extra fare, and generally handled B & O trains with the least possible dispatch. When faced with a B & O train, PRR dispatchers suddenly suffered spells of confusion and outright incompetence, holding the B & O for other trains that turned out to be nowhere nearby. Slow freights, switching moves, and light engines unaccountably would show up and occupy the track. PRR crews sometimes would decide to park their trains on the line and go off for lunch. Switches would fail to work properly, and switchmen would become preoccupied with other things. Instead of the six minutes normally needed to cover this section of track, it often took B & O trains anywhere from half an hour to two and one half hours to get through. Embarrassingly, one B & O train carrying members of Congress got this treatment and eventually arrived in New York five and one half hours late.

Another lawsuit followed, which was quickly decided in favor of the B & O-Reading-Jersey Central group. For the time being, B & O's New York runs were free of the Pennsy's hands altogether.

While Garrett was maneuvering to get his new Philadelphia-New York route, he also did what he could to speed up the slow, unwieldy PW & B connection in Baltimore. At the time, there seemed to be no way that B & O could build anything comparable to the Pennsy's direct line through the city. Baltimore's downtown and waterfront areas were heavily developed by 1880, and any new rail line on a private right of way promised to be extremely expensive. Indeed, the Pennsy's line was exactly that, even though it avoided the center of the city. So, as an expediency, the B & O and PW & B decided to transfer their traffic over water rather than through the streets. In 1879 the two railroads jointly ordered a trainferry to connect the two lines a short distance across the harbor. By using it, passenger trains could be kept intact and be quickly loaded and unloaded from the ferry.

Built by Harlan & Hollingsworth of Wilmington, the 324-foot-long *Canton* was a side-wheel steamer capable of carrying ten passenger cars or 27 freight cars. In June 1880 it began operating between the B & O's Locust Point marine terminal (on the south side of Baltimore harbor) and the PW & B's ferry slip off Boston Street in the Canton section of town. The new ferry route required B & O trains to use the Locust Point branch, which bypassed Camden station; as a result, they either had to make a stub-end backup move to reach Camden, or pick up and drop Baltimore cars south of the station. The operation obviously was no match for the direct all-rail PRR-PW & B service, but at least it avoided most of the PW & B's street trackage and, at long last, ended the Pratt Street horse transfer.

So, by the end of 1880, Garrett had a Pennsy-free Philadelphia-New York route and a marginally better way of getting his trains through Baltimore. B & O's December 19, 1880 timetable listed three through New York-Washington trains each way over the new B & O-Reading-CNJ route, averaging about six hours and fifty minutes to make the entire run. Through sleepers were carried between New York and Cincinnati, New York and Chicago, as well as New York-Baltimore, New York-Washington, and Philadelphia-Washington. All sleepers were advertised as B & O-owned fourteen-section "Palace Cars...the finest and most spacious ever constructed...finished in the highest artistic style." Luxurious equipment or not, the Chicago trip must have been a particularly grueling one. At that time B & O's Chicago trains followed a roundabout route through Grafton and Wheeling, West Virginia and Newark, Ohio and took 36-1/2 hours.

As the hectic events of the late 1870s were unfolding, something else was also happening to John Garrett. Although only in his late fifties, the normally robust Garrett had begun to change noticeably. Earlier, in 1873, he had suffered a physical breakdown attributed to strain, and took an extended trip to Europe to recover. But apparently there were other problems, too. William Keyser, a B & O vice president during this decade, later recorded that Garrett's mind began failing along with his body. In 1876 Garrett's doctor privately told Keyser that the B & O president was "suffering from brain fag." Keyser himself observed that "Garrett's physical condition had seriously impaired his mental vigor, which was, from time to time, conspicuously evidenced by his erratic and vacillating course, altogether foreign to his nature." Unfortunately, said Keyser, as his mind gradually deteriorated, "his willpower remained unimpaired, leading him to do, and persist in doing things which those around him advised against, and which ended disastrously."

In late 1879, in fact, Garrett's doctors sent him on another long European vacation. When Keyser visited him in Paris in the summer of 1880, he later noted that he "was much concerned to see how he had changed and aged since we parted." It was in this state that John Garrett fought the final battles of the New York-Washington route war.

By late 1880 events started moving quickly, but with odd zigzags. Although the PW & B still was reliably and neutrally handling both B & O and PRR trains between Baltimore and Philadelphia, Garrett suddenly veered off in another direction by joining a group which was planning a new Baltimore-Philadelphia railroad.

In Delaware was a twenty-mile short line called the Delaware Western, which started at Wilmington, twisted its way northwest through the scenic valley of Red Clay Creek, and ended at Landenberg, Pennsylvania just over the Delaware-Pennsylvania state line. Built between 1871 and 1872 as the Wilmington & Western, the little railroad originally was heading for Oxford, Pennsylvania and had vague hopes of forming part of a route from Wilmington to Pittsburgh. Such dreams died rapidly enough and the railroad never got beyond Landenberg, where it joined another railroad (later a PRR branch) running south from the Pennsy main line at Pomeroy, Pennsylvania. Left to live off the small on-line mills and quarries plus some interline coal movements through Landenberg, the Wilmington & Western went into receivership in 1875. In 1877 it was sold and reorganized as the Delaware Western.

Its new owner was Col. Henry S. McComb, a Wilmington entrepreneur and promoter who, among other things, had been an active railroad carpetbagger in the South after the war. At the time McComb bought the Delaware Western he was fresh from losing control of a railroad he had put together between New Orleans and Cairo, Illinois — later to be the Illinois Central's main line. The Delaware Western's charter allowed it to build eight miles in any direction from Wilmington, and McComb apparently saw it as the germ of a new railroad line between Philadelphia and Baltimore.

Somehow he got William H. Vanderbilt, the Commodore's son, interested in the idea. He also approached Garrett, who committed himself to a twenty percent interest and promised that the Jersey Central would pick up another fifteen percent. (Significantly, through all the maneuverings involving the New York-Washington business, Garrett always preferred to be merely a participant in financing syndicates rather than attempting full purchase or control himself.) McComb and his cohorts then submitted a bill in the Delaware legislature to broaden the Delaware Western's franchise rights.

At that point the PW & B realized that it was being surrounded by large, powerful forces which might build their own paralleling line or otherwise undercut its strategic position. Its management decided that it was time to sell to someone before the process went any farther.

The B & O obviously was one potential buyer. But once again Garrett was either unwilling or unable to deal strictly on his own. Perhaps this was because of his mental state, perhaps because he knew what the general public did not — that B & O's financial situation was becoming shaky and it could ill afford to take on any large new debt. For whatever reason, Garrett abandoned the Delaware Western syndicate and joined another investor group formed to

—THE—

BALTIMORE AND OHIO

IS THE ONLY LINE RUNNING

BETWEEN

CHICAGO AND NEW YORK

—VIA—

Washington, Baltimore and Philadelphia.

THERE BEING NO CHANGE OF CARS OF ANY CLASS

WITH ITS NEW AND PERFECT TERMINAL FACILITIES AT NEW YORK, THE BALTIMORE AND OHIO RAIL ROAD NOW POSSESSES ADVANTAGES OVER ALL OTHER LINES OUT OF CHICAGO.

2 TRAINS DAILY 2

BETWEEN CHICAGO AND THE EAST.

7.50 A. M., DAILY EXCEPT SUNDAY, AND 4.05 P. M., DAILY. THE 4.05 P. M. IS A SOLID TRAIN, RUNNING EVERY DAY IN THE WEEK AND ARRIVING AT WASHINGTON 9.45 P. M., BALTIMORE 11.05 P. M. NEXT DAY, PHILADELPHIA 3.05 A. M., NEW YORK 6.40 A. M. SECOND DAY.

The New York Terminus is but two squares from 9th Ave., and three squares from 6th Ave. Elevated Roads,

DOUBLE TRACK!
STONE BALLAST!
STEEL RAIL!
JANNEY COUPLER!

FOR TICKETS, RATES, SLEEPING CAR BERTHS, ETC. CALL AT No. 83 CLARK STREET, CHICAGO, ALSO OFFICES OF CONNECTING LINES THROUGHOUT THE WEST AND NORTHWEST.

From B & O's timetable of December 19, 1880. At this time the Chicago-New York cars were routed east through Newark, Ohio and Wheeling, West Virginia. B & O Museum Collection.

buy the PW & B, this one dominated by the notorious Jay Gould and including Russell Sage, August Belmont, and John Jacob Astor. Gould had previously bought into the Jersey Central and became a director in early 1881; reportedly he had vague plans to use the CNJ as an eastern terminal for his Wabash system. The Pennsylvania also was invited to take a minority interest, but declined.

In fact, for once the Pennsy stood on the sidelines. The expansionist era of Thomson and Scott already had ended. Scott, Thomson's survivor and successor, had resigned in 1880, debilitated by a stroke and frustrated by financial reverses and stockholder resistance. His successor, George B. Roberts, was neither a Thomson nor a Scott. A hard-working, competent civil engineer, Roberts was quiet, gentlemanly, and unaggressive. His more capable senior vice president, Alexander J. Cassatt, chafed at the inaction but for the moment sat still.

Most of the PW & B's stock was owned by Boston capitalists, and in February 1881 the Gould-Garrett group began negotiating with Nathaniel Thayer, one of the major Boston stockholders. Thayer, thinking he could control his fellow stockholders, committed himself to delivering a 51 percent controlling interest in the railroad for $70 a share. The deal was closed and announced on February 22nd; Thayer had until March 15th to round up the stock. Assured that the PW & B was theirs, Gould and Garrett went their separate ways.

What followed next became a legendary episode in eastern railroad history. Indeed, the legends have heavily shrouded the true events, making it difficult to sort out what really happened and when. According to one story, Garrett prematurely tipped his hand by visiting Roberts on March 6th, arrogantly announcing his coup, and graciously assuring him that the B & O would not interfere with the PRR-PW & B traffic. Other versions have him drinking too much and variously making the announcement at a dinner in Boston or at the Maryland Club in Baltimore. Whatever the truth in any of this, the Pennsy certainly knew what was going on. Both sides also knew something else — that some of the other PW & B stockholders were unhappy with Thayer's price and had quickly created a special committee to negotiate a better one. Garrett, however, did not take them seriously.

Prodded by Cassatt, who had been quietly talking to the Bostonians, Roberts suddenly but silently entered the fray. Cassatt and Roberts met the stockholder group March 7th, immediately bought a 39 percent interest at $78 a share, and offered the same price for any more stock delivered before April 1st. The Pennsy's eleven percent premium over the Gould-Garrett group's price was enthusiastically accepted, and it quickly gathered in 92 percent of the PW & B's stock.

Once again humiliated by the Pennsylvania, Garrett quickly decided to build his own Baltimore-Philadelphia line, whatever the cost. On March 23, 1881, only two weeks after the Pennsy's coup, B & O went back to the little Delaware Western, bought it outright, and prepared to head east on its own.

A DIFFICULT BIRTH

1881-1886

Having made his lightning decision to extend the B & O east of Baltimore, Garrett had to stop and decide how to do it — and especially how to pay for it. Time was pressing hard. Already the company had suffered the effects of past service interruptions, uncertainties, and a gradually diminishing presence on the New York route. Now the Pennsylvania had its own fully-controlled New York-Washington line, plus its direct service to the South and its quicker route through Baltimore, and it obviously was in a position to cancel B & O's joint agreements and services with the PW & B.

Unfortunately, more rate wars were waged in 1881 and 1882, holding down earnings and making any financing uncertain. Perhaps out of his gentlemanliness, however, Pennsy president Roberts allowed the PW & B to continue carrying B & O passenger trains and freight shipments although the end was all too clearly in sight.

By early 1883 things had become temporarily tranquil and Garrett could begin work. His ultimate goal was a B & O line all the way from Baltimore to New York, a double-track, high-speed railroad built to the highest pos-

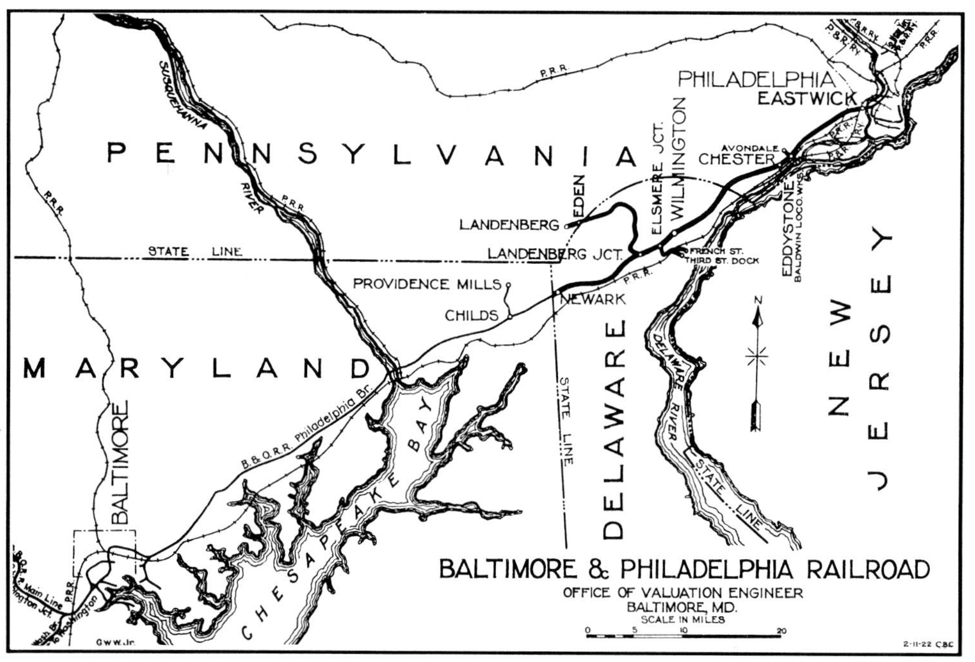

B & O's projected Philadelphia line was to be built under B & O's Maryland charter as far as the Delaware state line. The Baltimore & Philadelphia was a reincarnated and expanded version of the old Delaware Western, and was merely a "paper" company to own the charter rights through Delaware and Pennsylvania. The original Delaware Western (Wilmington & Western) began at the French Street and Third Street terminals in Wilmington and wiggled out to Landenberg. H. H. Harwood, Jr. Collection.

Piedmont topography demanded some Colorado-style bridges, such as this trestle over Big Northeast Creek, west of Childs, Maryland. Smithsonian Institution Collection.

sible standards. Ironically, he had persistently resisted spending much money to improve B & O's obsolete and woefully inefficient main lines west of Baltimore, which were far busier and more important. At the New York end he planned to solve the perennially annoying inconvenience of the Jersey City-Manhattan ferry transfer (a problem the Pennsy also shared) by using a large trainferry like the *Canton* to carry his passenger trains intact across the Hudson to a waterfront Manhattan terminal.

But the first step was to build between Baltimore and Philadelphia, where B & O could connect with the Reading-Jersey Central Bound Brook Route. B & O's original Maryland charter could take it as far as the Maryland-Delaware state line near Newark, Delaware. To build through Delaware and Pennsylvania, Garrett reincorporated his newly-acquired Delaware Western as the Baltimore & Philadelphia Railroad in February 1883. The Baltimore & Philadelphia's projected route ended on the west side of the Schuylkill River in Philadelphia, near the point where the PW & B met the Junction Railroad. The Junction, of course, was the key connecting link to reach the Reading's track several miles north in Fairmount Park. Unhappily, though, the Junction was now a dubious bet. When the Pennsy picked up the PW & B, it also inherited the PW & B's one-third share of the Junction, and now was a two-thirds owner.

To avoid the Junction Railroad, Garrett created still another subsidiary, the Schuylkill River East Side, which, as its name described, would build along the east bank of the river to the Reading. Roughly paralleling the Junction Railroad route, the "East Side" ran northward to Fairmount Park, joining the Reading's City branch south of Girard Avenue between 30th and 31st Streets, eventually to be called Park Junction. En route it would pass the site of B & O's Philadelphia passenger station at 24th and Chestnut Streets, some distance west of what was then the city's center. The East Side also would serve as B & O's Philadelphia freight terminal, with a projected branch across the far south side of town to reach piers and warehouses along the Delaware River.

To finance the Philadelphia line, Garrett managed to negotiate a fifty-year $11.6 million loan from Baring Brothers in London. With that, construction started in early 1883 at the Brandywine River in Wilmington.

By the early 1880s new railroad lines were being built with great frequency and amazing speed. The 514-mile Nickel Plate Road, for example, took less than two years to survey, build, and open its Buffalo-Chicago line. For B & O's Philadelphia extension, neither the ninety-mile distance nor the terrain were greatly challenging, and B & O announced that it would be in business by 1885. As it turned out, it would take close to four years to begin a stopgap operation and a full thirteen years to complete a fully competitive line. Along the way the railroad had to cope with every type of business problem: cost overruns, deteriorating finances, route and construction difficulties, commercial handicaps, political opposition, and even a crippling but unspoken human problem with the company's leadership.

The first and worst obstacle was getting out of Baltimore itself. B & O's line ended at Camden station, close by the harbor and the city's center; its primary freight terminal was at Locust Point, a peninsula on the south side of Baltimore harbor. To build eastward toward Philadelphia, it was blocked on one side by the densely-developed city and on the other by the harbor, which was too wide to bridge or tunnel at any reasonable cost. Ten years earlier the Pennsylvania had been faced with the same impasse when it invaded Baltimore with its Washington branch. The Pennsy found it necessary to build an extremely expensive route around the city's north side, digging four tunnels totaling two miles in length. Whatever route the B & O chose promised to be even more difficult, requiring either extensive elevated structures, tunnels, or both, and residents of the area were enthusiastic about neither.

Work under way on the Brandywine bridge in Wilmington, about 1885. Smithsonian Institution Collection.

The balance of the Philadelphia line was easier, but had its own obstacles here and there. Generally, the projected B & O route closely paralleled the old PW & B line, which had been built along mostly level Coastal Plain terrain. Yet although the B & O line was never more than four miles west of the PW & B (and usually closer), it often was on higher, hillier ground necessitating more curves and longer, higher bridges and trestles. The Susquehanna River was over a mile wide at the point the B & O planned to cross, and its line was to be over ninety feet above the river level. Smaller waterways, such as Big Northeast Creek in Maryland, the Brandywine in Delaware, and Darby Creek in Pennsylvania, required particularly large bridges.

Such problems would have been serious enough for a soundly-financed company like the Pennsylvania, but they gave Garrett very little maneuvering room. Through the later years of his long rule, Garrett had managed to keep up a good financial face, but it was increasingly an illusion. Earnings had been inflated by ignoring depreciation (a relatively common practice at the time), and the dangers of the ever-growing debt load were masked on the balance sheet by overvalued assets. To be blunt, the B & O really could not afford to build the line.

Commercially, the project promised to be an uphill battle, especially for freight business. To compete for freight traffic and set rates on the same basis as the Pennsylvania and other railroads, the B & O needed its own freight facilities and direct access to shippers and receivers. But as a latecomer in this already well-developed territory, it was forced to skirt, bypass, or squeeze in between existing communities, industries, and other rail lines.

At Philadelphia the Pennsy, the Reading, and their various predecessors had been busy since the early 1830s filling the city with a maze of main lines, branches, and industrial spurs. Philadelphia's industries already had grown up along PRR and Reading tracks, and there was no practical way that B & O could reach them on its own.

Although the Reading was a willing partner in B & O's passenger services, it had its own competitive freight routes to protect and thus would open few of its customers for B & O service. Neither was there much remaining space in Philadelphia to build new branches or locate new industry.

B & O was slightly better off at Wilmington. The old Delaware Western / Wilmington & Western's original main line gave it a two-forked branch to the city's downtown area and industry along the Christiana River. Here too, however, the PRR served most of the busy manufacturing plants. As for the B & O's main line through Wilmington, it followed the high ground about a mile west of the center of town, passing through semi-suburbs and not much else. In the hinterlands between Baltimore and Philadelphia, the B & O route avoided almost all the towns which the PW & B (now PRR) had been serving for the past 45 years. The only exception was Newark, Delaware, where B & O's line was actually located through the heart of town while the PW & B / PRR skirted it.

But all of that was secondary, however. Garrett's goal was direct participation in New York's city and port traffic, and its connections to New England. But at New York the problem was doubly difficult. B & O's own tracks would end at Philadelphia, at least for the immediate future. From there to New York it planned to continue using the Reading-Jersey Central route. This meant that B & O had no means of directly reaching intermediate freight customers anywhere between Philadelphia and New York; it would be limited to New York business alone. Garrett intended eventually to build his own line all the way to New York, but that final section had to await stronger finances.

At the same time, B & O did need its own freight terminals at New York. While its passenger trains still used the CNJ's Jersey City ferry terminal, the nature of the freight business demanded that B & O control and operate its own yards, piers, and marine services. But again, it was a latecomer looking for space. The New Jersey waterfront immediately opposite Manhattan already was either occupied or access was blocked by other rail lines. By 1883 six major railroads terminated on the west side of the Hudson, with several smaller companies using their facilities. The Jersey Central alone owned or controlled two and one half miles of Jersey City waterfront.

And on another front, B & O was all too aware of the political problems it faced — especially in Philadelphia, the Pennsylvania's seat of power. Symbolically, the Pennsy's grand new passenger terminal and headquarters office building, Broad Street Station, was directly across the street from Philadelphia's City Hall. Needless to say, the PRR had "friends" throughout the city's political structure. Even if it did not, B & O's role as the servant of Baltimore, Philadelphia's rival, would greatly dampen the railroad's welcome. Other potential political problems lurked along the New York-New Jersey waterfront.

Then there was the human problem. No sooner had B & O begun slogging into this quagmire than it lost its leadership.

Much of B & O's new Philadelphia line ran though barren territory. This later scene, taken about 1920 at Twin Oaks, Pennsylvania (west of Chester), was all too typical. B & O Museum Collection, B & O photograph.

Robert Garrett in happier days, looking every bit the country squire. His troubled reign over B & O lasted from 1884 to 1887. B & O Museum Collection.

B & O's New York extension was, of course, John W. Garrett's personal creation — an almost irrational act of spite and defiance after years of being bested by the Pennsylvania. But by the time construction started in 1883, Garrett was clearly sliding into the final stages of his physical and emotional decline. He had become increasingly reclusive, attempting to run the company from his home and using his wife as his go-between and guardian. When she was killed in a carriage accident in November 1883, Garrett went into a profound depression and died ten months later at age 64.

Garrett's oldest son Robert, 37 at the time, ran the company during much of his father's last year and officially inherited the presidency when the elder Garrett died September 26, 1884. It was no improvement. Unsuited and unprepared to take on John Garrett's mantle, Robert was characterized by one historian as "more a country squire than a railroad baron." More disturbingly, he seemed to suffer from his own physical and mental problems. Beset by business pressures (not the least of which was completing the New York line) and several personal shocks, he soon cracked. Much of his short presidency was spent traveling, attempting to escape and recover.

Fortunately the railroad's day-to-day management was largely in the hands of two stable and highly capable subordinates, vice president Samuel Spencer and general counsel John K. Cowen. Although both were greatly concerned over the B & O's worsening financial state, they gamely carried on the Philadelphia/New York project to Garrett's standards. At this point they had no choice.

To get into business as quickly as possible and to keep the initial costs reasonably low, B & O simply postponed the problem of getting across downtown Baltimore. Once again, it resorted to the expediency of a carferry across the harbor. As finally laid out, the new line to Philadelphia started on the east side of Baltimore's harbor in the Canton section, near Clinton Street and Danville Avenue. From there it headed directly northeast through mostly undeveloped land, crossing various PRR branches and the Pennsy main line at Bay View.

Construction was well under way in 1884 when the inevitable finally happened: in mid-May the Pennsylvania announced that it would terminate the old B & O-PW & B traffic contract on June 14th. The deadline was extended, but four and one half months later the PW & B notified the B & O that it would cease handling B & O's New York-Washington passenger trains October 12th. (Oddly, nothing was said or done about freight movements.) Knowing it was a losing battle, B & O went to court to stop or delay the action, but early in November it lost its case and gave up. Now, until it finished its own Philadelphia line, B & O was out of the passenger business east of Baltimore.

In the meantime the Pennsylvania began creating as many difficulties as it could within Philadelphia. Temporary trestlework suddenly appeared at the point where one PRR line crossed over the planned route of the Schuylkill River East Side. The Philadelphia City Solicitor also filed suit to prevent the East Side from occupying any city streets or installing a grade crossing at Gray's Ferry Avenue.

Between its legal problems and route location uncertainties, the East Side started construction later than the balance of the Baltimore-Philadelphia line, but finally got under way September 15, 1884. Its first section was a non-controversial stretch along the river between Cherry and Race Streets. But not much could be done on most of the line, since B & O faced a fight in getting its needed city ordinances.

In laying out its main line into Pennsylvania, B & O considered saving some construction by using the Reading's Philadelphia-Chester branch between those cities. This eleven-mile line began at Gray's Ferry in Philadelphia (where it joined the Junction Railroad) and ran to Chester, Pennsylvania, across the swampy lowlands near the Delaware River. It had been the PW & B's original main line, but the PW & B relocated its route to higher ground one and one half miles to the west in 1872 and leased the old line to the Reading in 1873. The idea was finally rejected after it was decided that the branch was not really suitable for B & O's main line use either and, additionally, would require crossing the PW & B main line at Chester.

The Reading Chester branch did however become an odd adjunct to the B & O line. Because of its PW & B heritage, it was not directly connected with any Reading track in the area; Reading traditionally had relied on the Junction Railroad to reach it from its main line at Belmont. With the Junction now two-thirds owned by the enemy, prudence dictated avoiding it altogether, so B & O and Reading worked out a trackage rights agreement allowing the Reading to use B & O's East Side line as its connecting link. Once the B & O line opened, Reading's Chester freights began operating over the East Side between Park Junction and Eastwick, at the west end of B & O's Schuylkill River bridge. The arrangement lasted through the Reading's corporate life and continues today with Conrail. In 1898 the Reading finally gave up all involvement in the Junction Railroad and sold its interest to the Pennsylvania, which then became the sole owner. This infamous line still exists in greatly altered form, with various segments now owned by Amtrak, SEPTA, and Conrail.

Predictably, getting into and through Philadelphia proved to be both a physical and political tangle. As much as possible, B & O's route avoided built-up areas, grade crossings, and public property. The East Side's main line hung close to the Schuylkill River bank, west of the city's center (which was then in the area between 4th and 9th Streets). Its planned freight branch to the Delaware River was a greater problem, however. The line would swing through mostly undeveloped land at the far south end of town, but would occupy and cross projected future streets at grade. But on both lines the Philadelphia politicians found locations to argue about: notably several street crossings west of the Schuylkill, the existing and future street use in South Philadelphia, and the area in and near Fairmount Park where the main line would join the Reading. (Among other things, the railroad would cross the park's main entrance.) Also, of course, the B & O had to cross the Schuylkill at a navigable point in the river, which meant dealing with marine interests and federal authorities.

It was not until May 1885, two years after initial work on the Philadelphia line started, that the B & O finally got around to the inevitable showdown over its routes within the city. For much of the next three months the railroad and city council were locked in arguments over street crossings, street alterations, and general line location. Throughout the scuffle, the B & O — particularly in the person of John Cowen — waged a subtle public relations campaign which included the anonymous distribution of a persuasive pamphlet (strongly attributed to Cowen) stressing the new horizons the B & O would open to Philadelphia and the blessings it would bring to the city.

Old Baltimoreans must have considered Cowen's rhetoric blasphemous, but it helped him to get his way. By mid-July the ordinances allowing B & O's routes through the city were passed and most other major legal hurdles were cleared. Already a year behind schedule, B & O's Philadelphia construction contractors went to work. Among other changes from its original plan, B & O substituted a tunnel for a planned elevated line at the Fairmount Park end of its route near Park Junction. (Today the tunnel passes close to the Philadelphia Museum of Art.)

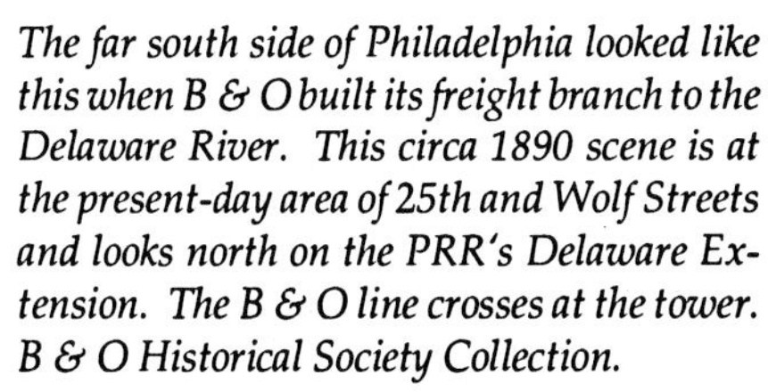

The far south side of Philadelphia looked like this when B & O built its freight branch to the Delaware River. This circa 1890 scene is at the present-day area of 25th and Wolf Streets and looks north on the PRR's Delaware Extension. The B & O line crosses at the tower. B & O Historical Society Collection.

Money also became a problem. By this time costs were far exceeding the original $11.6 million loan negotiated in 1883. In November 1885 an additional $4.5 million bond issue was floated to finance the Philadelphia work which, as it ultimately turned out, still would not cover everything. In addition, the railroad eased the strain on its treasury by deliberately dragging its feet in litigation over the value of property it had taken; some of these property payments, which totaled perhaps $2 million, were not made until two or three years after the line opened.

While fighting its Philadelphia battles, B & O also had to find an available and cheap location for its New York harbor freight terminal and even in the mid-1880s, New York was anything but a low-cost area. Its solution was odd, imaginative, and again, expensive: Staten Island. Located on the southwest side of the harbor, the island was still a large rural expanse, physically isolated from its surroundings by the harbor and by the Arthur Kill, a navigable waterway which separated it from New Jersey. Ferries connected the island with lower Manhattan and with Perth Amboy, New Jersey, but neither a road nor railroad reached it directly. A scattering of water-served industries had appeared along its shoreline, but for the most part its waterfront was undeveloped. A local passenger-oriented railroad ran twenty miles down the length of the island from Clifton, the Manhattan ferry terminal, to Tottenville, opposite Perth Amboy on the Arthur Kill. Completed in 1860, the little Staten Island Railway had been partly financed by one of the island's natives, an enterprising steamship and ferry operator named Cornelius Van der Bilt — as his Dutch name was originally spelled. With no outside connections except by water, the railway existed mostly to move people between the island's tiny towns and to the ferries at Clifton and Tottenville.

After an 1871 disaster involving one of its New York ferries, the railway was sold and reorganized, although the Vanderbilts maintained an interest in it. Erastus Wiman, a transplanted Canadian who had come to New York in 1867

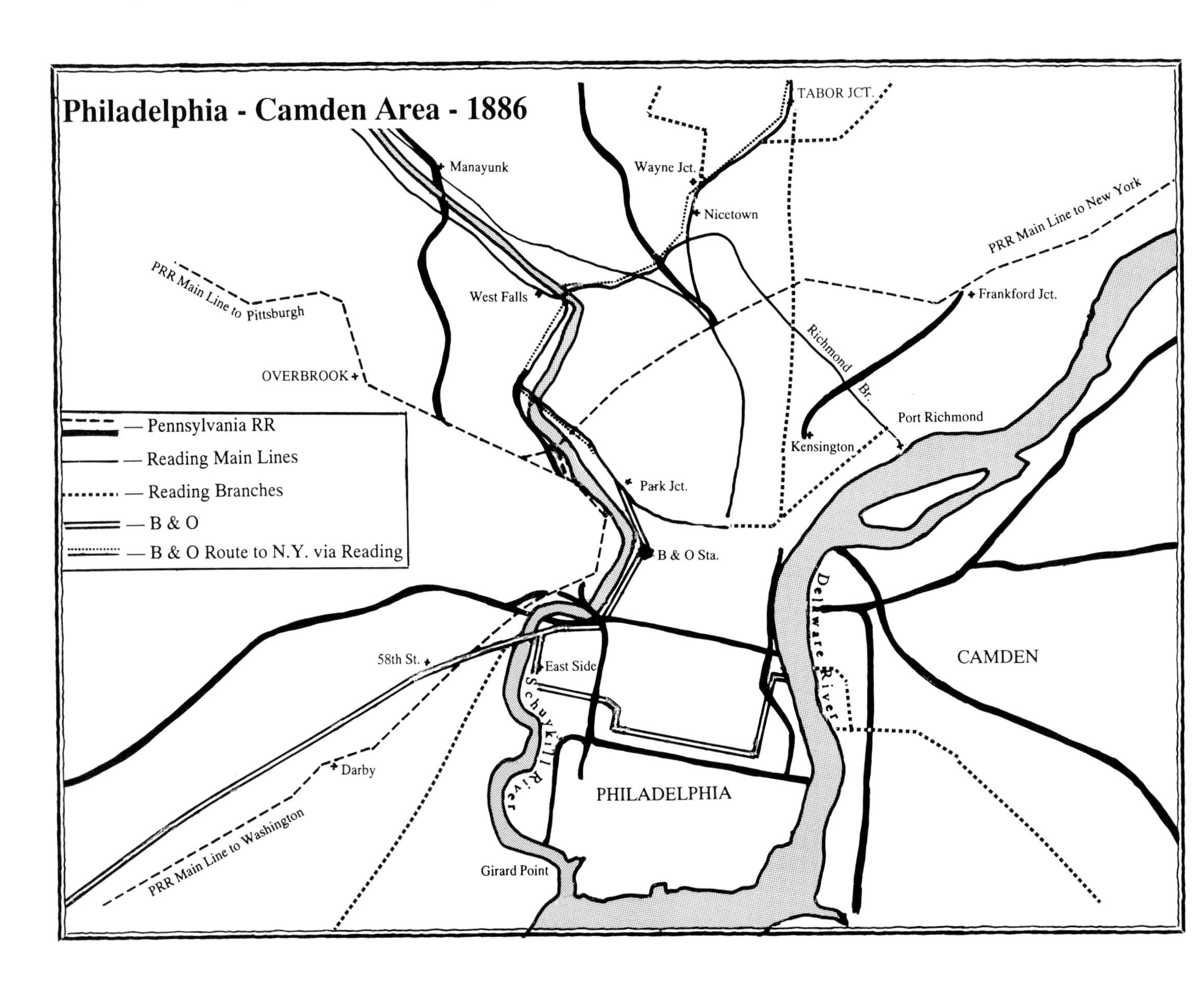

as R. G. Dun & Company's representative, settled on Staten Island and got interested in the railway more or less as a civic venture. In 1883 Wiman developed a plan to extend the line to St. George, one and one half miles north of Clifton at the northeast tip of the island, a more favorable spot for a ferry terminal. Then from St. George he planned to swing westward along the island's north shore, perhaps continuing all the way to mainland New Jersey to join some other railroad.

William H. Vanderbilt, Commodore Cornelius's son and the heir to his railroad empire, also saw Staten Island as a strategic railroad terminal, although not for his own system. An ally of the Reading's Franklin Gowen (particularly in their mutual warfare with the Pennsylvania), Vanderbilt originally had suggested that the Reading might make use of it. That idea died, but Wiman contacted Robert Garrett and proposed that he look at the island for B & O's freight terminal. Garrett did, and was interested.

In 1880 Wiman had incorporated a new company, the Staten Island Rapid Transit, to build his hoped-for extensions. In 1884, now with B & O backing, the SIRT leased the old Staten Island Railway and started construction of its own line. B & O's purchase negotiations took longer, but in November 1885 Garrett acquired control of the SIRT.

The B & O's original plan was to continue the SIRT extension west along Staten Island's north shore to Arlington, at the northwest corner. It would then cross the Arthur Kill into New Jersey and aim southwest to meet the Reading at Bound Brook. Thus B & O freight for New York would be hauled by the B & O to Philadelphia and the Reading from Philadelphia to Bound Brook; there the SIRT would take it to the Staten Island piers and carfloats. Although B & O passenger services would use the CNJ from Bound Brook to Jersey City, it was thought best to avoid the Jersey Central for freight. The early plan did, however, include a CNJ connection at Cranford, New Jersey.

The B & O-Staten Island purchase deal was barely concluded when the ill-starred Robert Garrett got the first of several emotional blows which helped end his business career and eventually his life. On December 8, 1885, he called on William H. Vanderbilt at his New York home, apparently to discuss some details of B & O's Staten Island and New York harbor situation. Midway through the private meeting, the hypertensive Vanderbilt dropped dead of a stroke before Garrett's terrified eyes.

The Staten Island project had a few problems of the Philadelphia variety, too. The Arthur Kill was (and still is) a busy waterway connecting portions of lower New York harbor, and B & O's bridge would require Congressional approval. As at Philadelphia, it planned a conventional swing bridge. Predictably, the marine and pier operators in the area objected — particularly the Pennsylvania and the Lehigh Valley railroads, both of which had waterfront

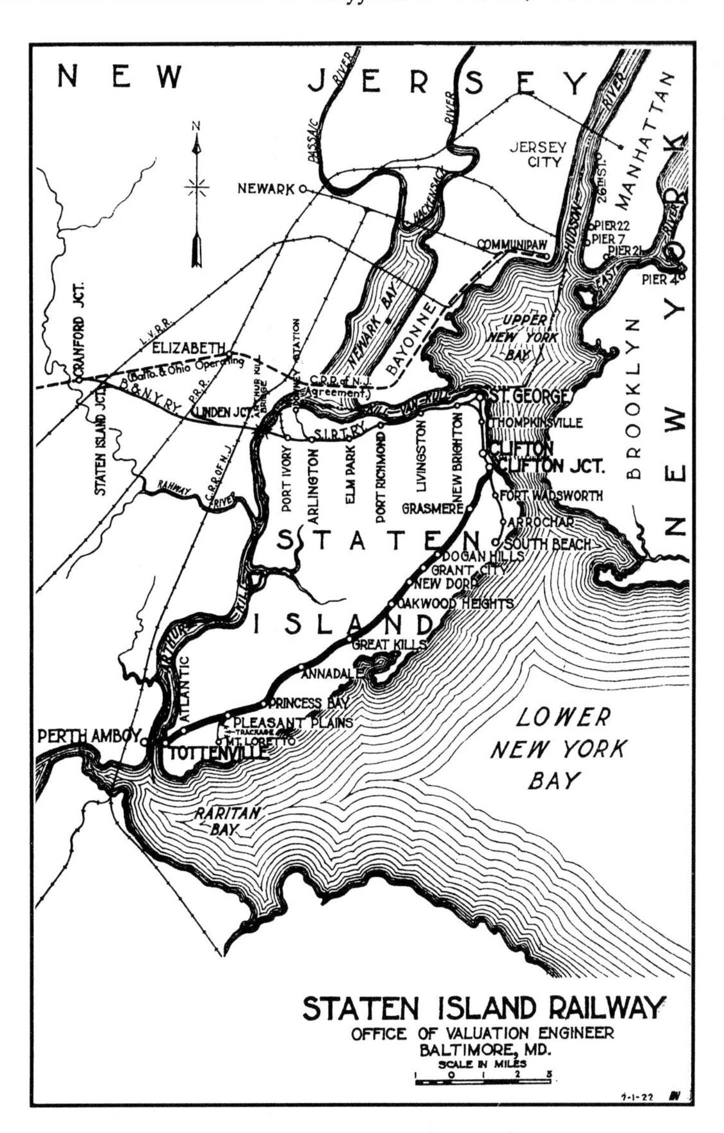

The original Staten Island Railway route from Clifton to Tottenville is shown by the heavy line. A lighter line traces the newer B & O-financed construction to St. George and into New Jersey, ending at Cranford Junction at the left on this map. The dashed line is the Jersey Central's main line to Jersey City (labeled "Communipaw" at the upper right). H. H. Harwood, Jr. Collection.

coal terminals at the Amboys, south of B & O's proposed crossing. The water route also was used to reach the PRR-owned Delaware & Raritan Canal on the Raritan River.

Nonetheless, construction of the Staten Island freight connection went on concurrently with the last stages of the Baltimore-Philadelphia line work. In 1886 the SIRT was completed from Clifton through St. George to Arlington, at the east end of the proposed Arthur Kill bridge. But the remainder of the route, although short, was destined to be delayed longer, adding yet another financial strain on the wobbling young Garrett. The legal fight over the Arthur Kill bridge took almost one and one half years to resolve; actual construction of it did not start until July 1887. Other problems delayed the connecting railroad line in New Jer-

First class standards show clearly in this 1920 view at Stepney, Maryland, a tiny hamlet west of Aberdeen. B & O Museum Collection, B & O photograph.

sey, including a major route change. Work on this section did not even start until 1889, over two years after the Baltimore-Philadelphia line had opened and through New York service had begun.

Although the Staten Island connection was still in the future in 1886, the Baltimore-Philadelphia line was at last getting close to completion. With most basic construction work finished (except at the Philadelphia end), B & O

Rossville, Maryland, seven miles east of Baltimore, was idyllically rural when B & O built through. Now a semi-industrial suburb, it looked like this at the turn of the century. Smithsonian Institution Collection.

This iron truss swing bridge carried B & O's line across the Schuylkill River in Philadelphia. Smithsonian Institution Collection.

turned its attention to stations and other facilities. In carrying out John Garrett's dictum to create a first-class railroad, it paid particular attention to the appearance of its stations. The job of designing the most important structures went to Philadelphia's most outstanding architect of the time, Frank Furness. Then in his mid-forties, the eccentric and highly individualistic Furness was at the peak of his career and his creative powers. "His were among the most boisterous and challenging buildings in an age and a city noted for aggressive architecture" said architectural historian James O'Gorman. And indeed, Furness combined the exuberance of the Victorian era with originality and whimsy. B & O commissioned him to do the Philadelphia station at 24th and Chestnut Streets, the station at Chester, Pennsylvania, and two at Wilmington: the main line station at Delaware Avenue and DuPont Street and a smaller downtown terminal at Market and Front Streets, at the end of the old Delaware Western/Wilmington & Western line.

In addition to these large stations, the railroad planned at least 33 smaller passenger or passenger-freight stations between Baltimore and Philadelphia, an average of one station every two or three miles. At the Philadelphia end, many of these were spaced less than a mile apart for the convenience of potential commuters. Since the B & O route skirted most population centers, many of these 33 stations would serve little more than hamlets or roads leading to the center of some town. Nevertheless, all of them were to be substantial and attractive buildings, designed by professional architects rather than railroad draftsmen.

As the new railroad took shape, the B & O also formalized agreements with the Reading and the Jersey Central for handling business between Philadelphia and New York. Made effective July 27, 1886, the agreements in effect spelled out a three-way partnership, with each of the three railroads responsible for its own part of the haul and with revenues split among them based on mileage. Both freight and passenger traffic were covered by the same basic contracts, although details differed. For the through New York-Washington passenger services, each railroad would supply a proportion of the needed cars, again roughly based on the mileage of their part of the route. Since the Jersey Central's portion of the haul was only thirty miles but included expensive terminal services at Jersey City, it was allowed extra handling charges for both freight and passenger traffic, and was not required to contribute passenger equipment unless it chose to. B & O was expected to supply all cars and personnel for its package express services handled on passenger trains.

In railroad jargon, the B & O-Reading-CNJ arrangement was a joint traffic agreement rather than a trackage rights contract. Technically, the contract made each railroad an independent partner, managing its own part of the operation. Reading and Jersey Central locomotives and crews would handle B & O trains between Philadelphia and Jersey City; these were, in fact, really Reading and Jersey Central trains. In future years the legal arrangements for operating B & O's New York passenger trains would take several other forms, but its freight business always moved under variations of the original 1886 traffic agreement.

By mid-1886 the Philadelphia line was far enough along to allow some limited operations. On May 11th a three-car special train left Camden station in Baltimore for the first official trip. Aboard were Robert Garrett, Samuel Spencer, John Cowen, and assorted other railroad officials and contractors. Crossing the spectacularly long and high Susquehanna River bridge was a special event; the bridge was decorated with American flags, and a one-car train filled with workmen preceded the presidential special across — presumably to be a human sacrifice in case the bridge failed. The special itself took half an hour to cross the mile-long structure, amid much whistle-blowing. The arrival in Philadelphia was an anticlimax, however. Contractors still were feverishly working on the Schuylkill River bridge and the East Side line, so the train was turned at a temporary terminal at 58th Street, west of the river and far from the city center.

Baltimore-Wilmington freight operations began May 25th; by July 11th the Schuylkill bridge was ready and freights started running into Philadelphia itself. The difficult tunneling near the Reading connection at Fairmount

B & O No. 233, an 1870 Davis camel, pauses somewhere on the Philadelphia waterfront in 1887. H. H. Harwood, Jr. Collection, Smith photograph.

Park was far from finished, so any New York freight shipments had to be moved to the Reading over the old Junction Railroad connection. Whatever the problems, B & O had no choice. Although the Pennsy had evicted B & O passenger trains from the PW & B almost two years earlier, it had continued to handle its freight. But it finally cut the last ties and notified the B & O that the freight agreement would be canceled August 10th. B & O's first Philadelphia Division operating timetable, dated July 11, 1886, showed two through freights and one local freight each way between the Canton pier in Baltimore and the temporary Reading/Junction Railroad connection at Eastwick.

After some difficulty, passenger services began to appear in the early fall. On August 23rd a Baltimore-Wilmington local service started, and on September 19th B & O ran its first Baltimore-Philadelphia trains, three express runs each way plus a local. But as yet there was no direct connection with the Reading, so the passengers for New York or other points east of Philadelphia had to transfer across downtown Philadelphia to the Reading's station at 9th and Green Streets. (Starting August 30th, package express shipments and some immigrant traffic were routed to and from New York over a hideously roundabout route via Wilmington, Birdsboro, Pennsylvania, and Wayne Junction in Philadelphia — a 110-mile detour to cover a direct distance of about 33 miles.)

Some freight did manage to struggle through to New York, and in October the B & O officially went into the harbor floating business, temporarily using Jersey City as its terminal. Six carfloats, each holding ten cars, were bought and tugs were hired as needed.

At last, on December 7th, the Reading connection at Park Junction was completed. A week later, on December

Park Junction in Philadelphia became the eastern end of the B & O system in 1886; afterwards, all system mileages were measured from this point. This early 20th Century view looks northwest on the B & O line toward the junction. The Reading's City Branch is at the right and Fairmount Park is to the left. Smithsonian Institution Collection.

15th, the first B & O train in over two years ran through between Washington, Baltimore, and New York.

So by the end of 1886, B & O's New York extension was at least operational, but it was far from finished. Many of the intermediate stations were uncompleted or not yet begun and, as of late 1887, thirty miles of the line still were single track. There was no firm plan to solve the Baltimore harbor dilemma, where the carferry transfer would hamper schedules for over eight more years. Because of all the delays at the Philadelphia end, it had been decided to build a temporary ten-stall wooden roundhouse remotely located west of the Schuylkill at 58th Street, three miles from the main passenger station. And although the B & O was at last physically joined to the Reading at Park Junction, the Reading's own Philadelphia trackage was not laid out in a way to allow a continuous direct movement from there to the Bound Brook-New York line. As a result, B & O trains spent half an hour being backed through town to cover the six miles between the B & O Chestnut Street station and the Bound Brook route at Wayne Junction.

And finally, because of the late start at the New York end, work was only beginning on the freight line to Staten Island, which was to be B & O's own harbor terminal for New York business. As a stopgap, B & O freight was handled through the Jersey Central's Jersey City facilities under the interline traffic agreement.

But at least Garrett's dream now had solid form. From here on it was up to his successors to make it into a viable competitor and find the money to sustain it.

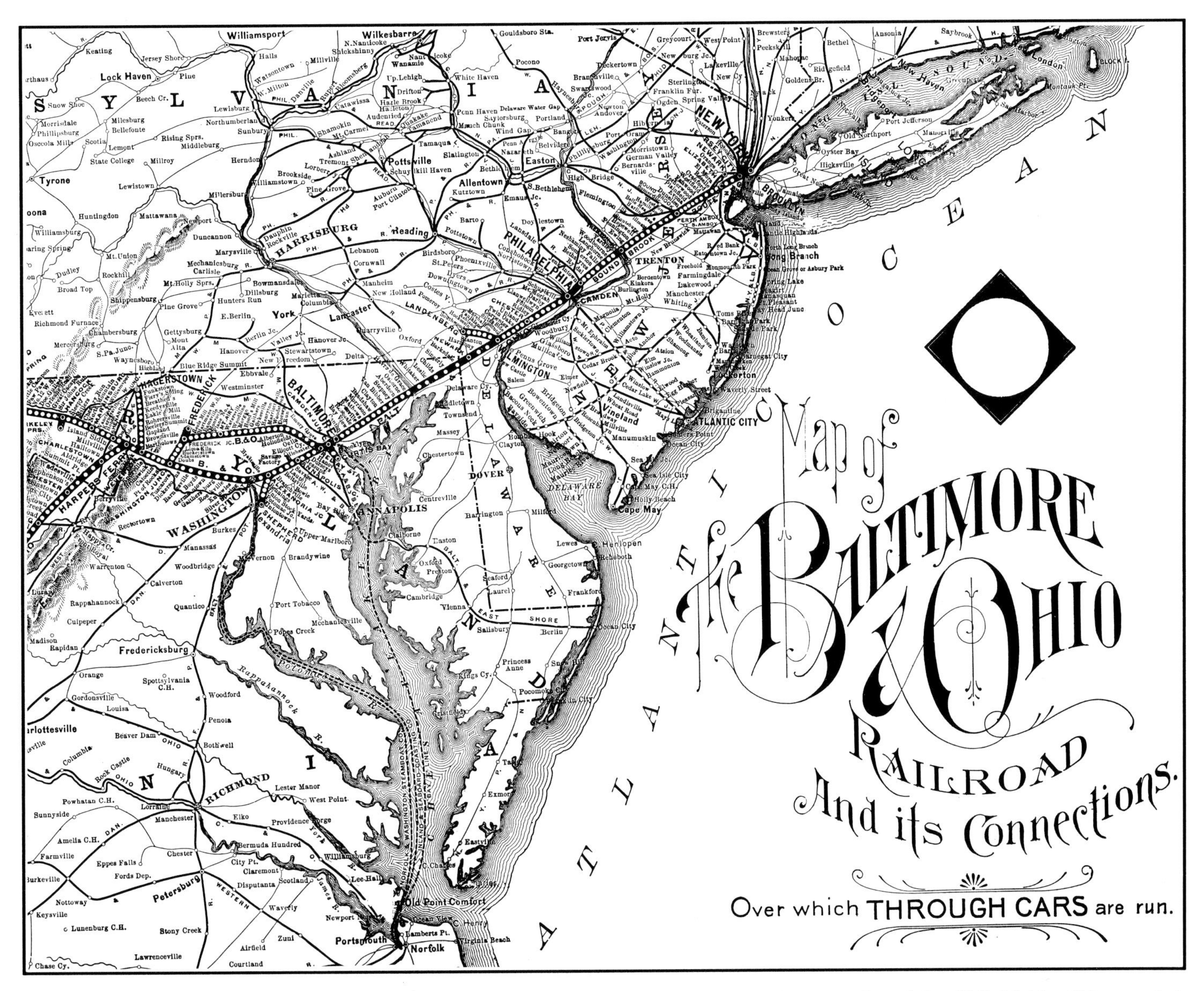

After 1886 B & O timetables could again show a direct line to New York, although the railroad's own tracks ended at Philadelphia. This example is from 1893. E. L. Thompson Collection, B & O Historical Society.

JOHN GARRETT'S RAILROAD 1887-1890

From 1887 to 1890, the B & O had to scramble to put its new line into final shape while at the same time trying to rebuild its badly bruised New York business.

Physically it was an excellent piece of railroad which was exactly what John Garrett had ordered. In fact, together with B & O's Washington branch, the Philadelphia line had the highest engineering standards of any part of the railroad's system. Built through rolling terrain and running crosswise to the drainage patterns, it had numerous curves and dips. But they were gentle since no curve exceeded eight degrees and most were less. Except for a

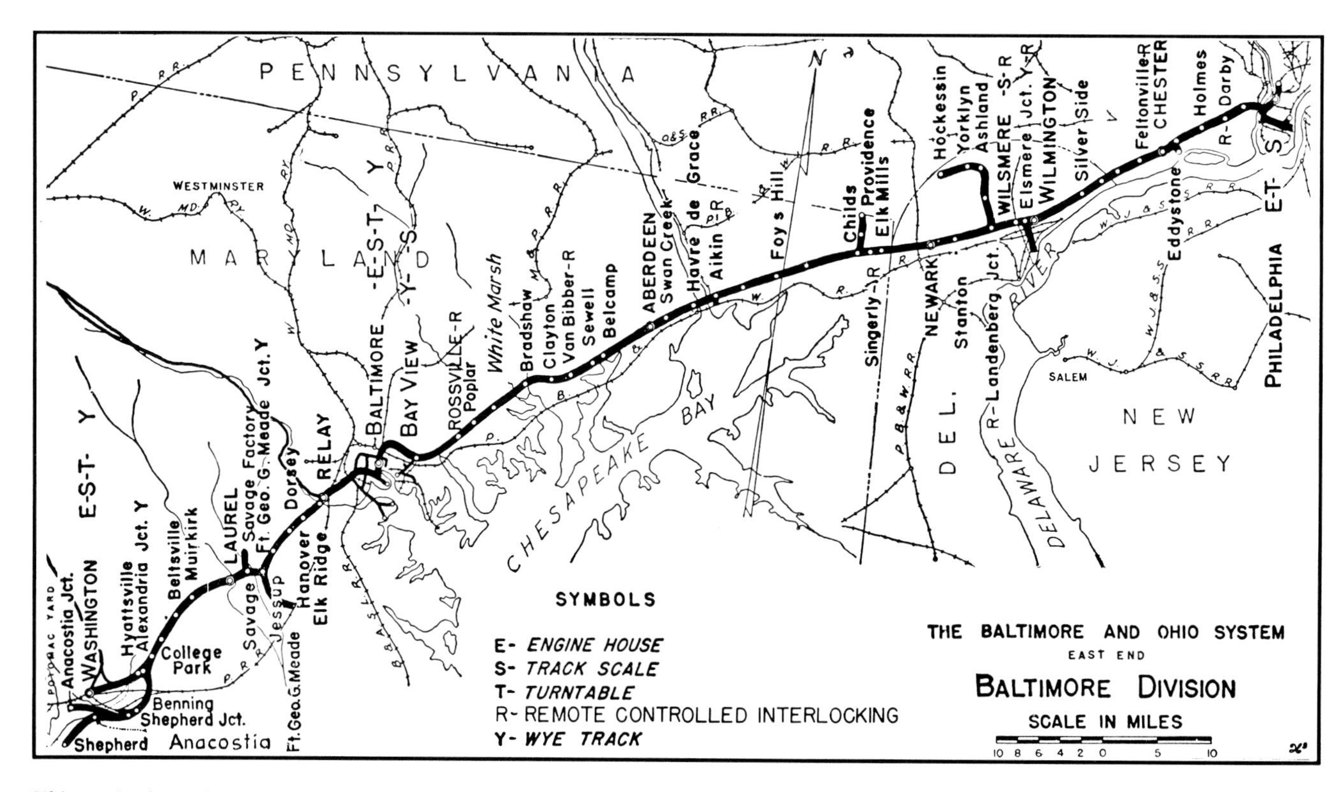

This semi-schematic operating timetable map dates to the 1940s. The paralleling line labeled "PB & W" actually is the PRR main line, made to look crookeder than B & O's — and also made to look like it goes nowhere. H. H. Harwood, Jr. Collection.

The newly-completed 1.2-mile-long Susquehanna River bridge. The camera looks northwest from Havre de Grace, Maryland. Palmer's Island, now called Garrett's Island, is at the right center. Smithsonian Institution Collection.

short westbound climb of 1.44 percent to get out of the Schuylkill valley between Eastwick and 58th Street in Philadelphia, grades were no more than 0.8 percent and seldom were sustained for more than one or two miles.

The Baltimore-Philadelphia line had two "summits" — one at Silverside, Delaware, between Chester and Wilmington (227.4 feet above sea level), the other at a spot four and one half miles east of the Susquehanna River called Foy's Hill (at 217.7 feet). Otherwise, it was largely an up-and-down profile, particularly in the hilly section between Baltimore and Van Bibber, Maryland (near Edgewood), which had a constant succession of short 0.7 to 0.8 percent climbs and drops.

Many bridges were needed to cross creeks, rivers, roads, and other railroads (primarily the PW & B and its branches), all of them designed for high speeds and what were considered to be heavy loadings. The smallest bridges and culverts usually were of stone; all others were iron, mostly pin-connected Pratt trusses (in both deck and through varieties) or girders. The railroad generally was located on higher ground than the older PW & B line and many waterways had to be crossed at points where they had carved deep, wide valleys — demanding large truss spans or iron trestles.

The crown jewel of bridges, of course, was the Susquehanna River crossing, which combined samples of almost every design type. The structure was 6,346 feet long and soared 94 feet above the low tide level. Midway across the river it hurdled an uninhabited wooded island originally called Palmer's Island and, by legend, the site of the first European settlement in this area. (Palmer's Island is now known as Garrett's Island, in honor of John Garrett. According to another legend, Garrett intended to build a resort hotel on it; at the time, the area was a fashionable retreat for well-heeled duck hunters.) Thus the bridge had to cross two separate river channels and a land area which it did with seven long deck truss spans, two large through trusses, and 2,288 feet of iron trestlework. In deference to its $1.7 million cost, the Susquehanna bridge was built to carry only a single track — the only single track section on the line. It (and later its replacement) would be the longest continuous bridges on the entire B & O system.

Far shorter, but spectacular in its own way, was the Brandywine Creek bridge at Wilmington. Its six deck truss

Wilmington's Brandywine bridge as completed, looking west. Bypassed by the railroad in 1910, the structure survived until 1980. Smithsonian Institution Collection.

Boone tunnel was essentially unchanged in this late 1930s photograph as a westbound Q-4-powered freight erupts. D. A. Somerville photograph.

spans totaled over 800 feet in length, and five high stone piers lifted it 110 feet over the river valley. Equally impressive was a long, high combination trestle and triple-span truss bridge over Darby Creek in Pennsylvania, and a spindly, somewhat frightening-looking iron trestle over the Big Northeast Creek east of Leslie, Maryland. The Schuylkill River was crossed on a low, four-span through truss bridge with a center swing section.

Frank Furness's inventive Philadelphia station opened onto the south side of Chestnut Street at the Schuylkill River. The main line tracks passed below the street at about the camera's location. B & O Historical Society Collection, B & O photograph.

The line also included three tunnels, all of them at the Philadelphia end. A short 626.5-foot cut-and-cover tunnel sliced through a hill in suburban Boone, Pennsylvania, west of Darby, and an even shorter one burrowed under busy Grays Ferry Avenue and the old PW & B main line in South Philadelphia. The longest and most difficult was a 2,303-foot cut-and-cover bore through a section of Fairmount Park as the line approached the Reading connection at Park Junction. In 1921 the Fairmount Park tunnel was partially rebuilt and lengthened to 2,540 feet.

The Philadelphia line's stations were in a class by themselves, notable then and now considered lost treasures (although not necessarily highly regarded in between). Frank Furness's Philadelphia terminal at 24th and Chestnut Streets is considered one of his finer works. It had a two-level layout, with entrance and facade facing Chestnut Street and tracks below street level along the river. Passengers entered a large waiting room with a massive fireplace, flanked by a ladies' room, restaurant, lunchroom, and barbershop. Stairways led down to the train concourse, ticket office, and baggage room. To the rear, stretching south from the station, was a rather ordinary iron trainshed 300 feet long and 100 feet wide spanning four stub tracks. The station's brick exterior was a riotous hodgepodge of shapes, which included a fanciful two-stage clock tower, dormers, gables, top-heavy tapered chimneys, arched windows, and ornamental ironwork.

Furness's station for Chester, Pennsylvania, was inevitably smaller and more subdued, but only by comparison. Built of brick and wood, it had many of the same elements as the Chestnut Street station — gables, dormers, tapered chimneys, and elaborate curved iron chimney braces, all of which led up to a central clock tower topped by an ornamental iron weathervane. During the station's last years the clock tower was amputated, leaving it still picturesque but unfocused.

The street-level waiting room included an odd juxtaposition of exposed steel girders contrasted against Victorian touches such as the ornate fireplace and, at the left, the restaurant entrance daintily decorated with intricately leaded stained glass. B & O Museum Collection, B & O photograph.

Downstairs at the track level was this spartan waiting room and ticket office. B & O Museum Collection, B & O photograph.

At the station's rear was this four-track trainshed. This 1946 photograph was taken from the Chestnut Street bridge over the river. L. W. Rice photograph.

Chester, Pennsylvania at the turn of the century. This area is now surrounded by Interstate 95. Smithsonian Institution Collection.

The main line station at Delaware Avenue, Wilmington, was ponderous but full of different shapes and forms. Smithsonian Institution Collection.

The Newark, Delaware station, shortly before its demolition in 1946. B & O Historical Society Collection.

***The photos on this page and on the opposite page** show a sampling of the "standardized" stations on the new Philadelphia line. In the photo above, the exotic design, which was repeated at four locations, is at Jackson, Maryland. It probably was a Frank Furness creation. Smithsonian Institution Collection.*

Bradshaw, Maryland, a design used for four stations on the line. Smithsonian Institution Collection.

Wilmington's two Furness stations were different from each other and from everything else on the line. The main line station at Delaware Avenue — another two-level design — gave a quick first impression of a ponderous stone fortress. But on further examination it revealed another eccentric collection of diverse shapes and odd details. A two-story brick and wood station downtown near the foot of Market Street served local trains of the old Delaware Western/Wilmington & Western line to Landenberg, as well as trains of the Wilmington & Northern (later absorbed by the Reading) and contained local company offices. The Newark, Delaware station, unconfirmed but also thought to be by Furness, was of yet a different design with gables, curved brackets, arched walkways around the building, and a central cupola.

Twin Oaks, Pennsylvania and four other locations received examples of this ornate residential-style station design. Smithsonian Institution Collection.

The countryside between Baltimore and Philadelphia was mostly unspoiled but also unscenic, a succession of woods, fields, and small settlements. But the frequent new local stations helped interrupt the blandness. From the largest to the smallest, they were all late-Victorian jewels. Although B & O wanted each of the 33 intermediate stations to be distinctive, it compromised with the cost and the admittedly low traffic potential of many of these spots. Eight more-or-less standardized station designs were drawn up and each used at anywhere from two to five different locations. But although "standardized," the stations hardly seemed so to travelers or the inhabitants of the tiny communities. All were architect-designed and anything but ordinary. Each type was notably distinct from any of the others, and there were enough different types so that they could be placed to give an impression of variety as one went along the line.

Three of the "standard" types were all-wooden buildings ranging in size from an elfin style used at such Maryland stops as Frenchtown (now Aiken) and Baldwin (later Elk Mills) to an elaborate, gingerbready two-story design for Holmes and Twin Oaks, Pennsylvania and three other locations. The other five varieties were wood and brick combinations, mostly single-story or one and one half-story buildings. The "typical" Philadelphia Division station had agent's living quarters, either upstairs or on a single floor. Frank Furness designed the Frenchtown (Aiken) station and, although records are lacking, there is strong evidence that he was responsible for as many as five of the eight "standard" designs. Another one of the eight, used at Boothwyn and Felton (Feltonville), Pennsylvania, was done by Baltimore architect, A. H. Bieler.

Of all the local station styles, the most picturesque — and most Furness-like — was a group of four free-spirited structures built in Maryland at Van Bibber, Havre de Grace,

Four Philadelphia-area suburban stations, such as this one at 60th Street, were built in this style. Smithsonian Institution Collection.

Whitaker (later Jackson), and Singerly. With high peaked roofs, ornamental dormers in several shapes, tapered chimneys, flared siding with "fish scale"-shingled woodwork, and ornate curved brackets, these had the same exuberant, offbeat feeling as Furness's small Reading stations built several years before.

Built strictly to reach Philadelphia and New York, B & O's new Philadelphia Division had minimal branches and would never develop many more. The only branch of any length was the old Delaware Western/Wilmington & Western line, which B & O had bought strictly to get itself charter rights in Delaware and Pennsylvania. The DW's route started as two separate branches on either side of the Christiana River in the south end of Wilmington; it then wandered northwest to Landenberg, about nineteen miles. Only a short 1.3-mile segment of its route outside Wilmington (in the area of what is now Wilsmere Yard) was used for B & O's main line; the balance, in effect, became two branches extending on either side of the main line.

The south section, called the Market Street branch, left the main line at East Junction (later Elsmere Junction) and ran three miles into Wilmington, closely paralleling the Wilmington & Northern. (The W & N was constructed as an independent line running from Wilmington to Reading, Pennsylvania, generally following the Brandywine north of Wilmington. In 1898 it was bought by the Reading.) The Market Street branch was essentially an industrial line and gave the B & O one of its few opportunities to reach established industries anywhere in the Philadelphia-Wilmington area.

The other end of the Delaware Western became B & O's Landenberg branch, winding fourteen picturesque miles from the main line at West Junction (later Landenberg Junction) to Landenberg, where it joined the Pennsylvania's Pomeroy branch. Never more than a light-density short line, it served a succession of small Red Clay Creek mills at points such as Greenbank, Faulkland, Wooddale, Ashland, and Yorklyn. Kaolin quarries near Hockessin, Delaware also supplied some traffic. If nothing else, the Landenberg branch was probably the prettiest piece of railroad on B & O's Philadelphia Division.

An infinitely more important branch was B & O's hard-fought line across the far south side of Philadelphia to the Delaware River. As originally built, it dropped directly south from the main line at the east end of the Schuylkill

Frank Furness designed this elfin structure for Frenchtown, Maryland, later renamed Aikin (Aiken), shown here in 1892. Smithsonian Institution Collection.

River bridge, then turned and worked its way east to the waterfront over Wolf Street, 23rd Street, and Oregon Avenue. At the time the branch was constructed, this part of the city still was mostly open fields, and the streets had not been opened. At Meadow Street the branch swung north again to the river at Dickinson Street, with several spurs to piers in the area. Over the next several years various waterfront facilities were built, including a coal pier at Jackson Street in 1893. Several piers also were built or leased north of the point where B & O's tracks ended, which were reached by tugs and carfloats. The Pennsy's Delaware Extension ran through much the same area as B & O's branch, and the two railroads crisscrossed each other at grade at five points. B & O's Delaware branch originally was intended to tap the city's port commerce; but in addition it became the railroad's only means of developing future on-line freight customers in Philadelphia, since its main line through the city was poorly located for industry.

The Leiper line, a freight branch barely visible on a map, was acquired shortly after the Philadelphia line opened. In March 1887 B & O bought a 2.5-mile quarry tramway running along Crum Creek just east of Chester, Pennsylvania. This primitive horse-powered operation had been built in 1852 by the quarry owner, George Leiper, to haul stone from the quarry site at Avondale to a dock on the creek near the Delaware River. Near the south end of its route it crossed under the new B & O main line near the B & O's Leiperville station (which in 1888 was renamed Fairview, then later Woodlyn, and finally Eddystone). The little tramway had historical distinction of a sort, since a short section at the quarry end was laid on the bed of the original 1809 Thomas Leiper tramway, reputedly the country's second earliest railway.

More importantly, the Leiper line gave B & O a much-needed freight traffic source, and it was immediately rebuilt for steam railroad operation. Although the quarry traffic eventually dried up (the branch's upper end was abandoned in 1942 after long disuse) the old tramway's lower end began to come alive after 1906 when the Baldwin Locomotive Works started development of its new Eddystone plant near the Leiper landing. By 1928 Baldwin had moved its entire production to Eddystone in a completely modern locomotive-building facility which sprawled over 591 acres and included ninety buildings. Completed almost on the eve of the Depression and, worse, the dawn of the diesel locomotive, the Eddystone plant seldom worked at its full capacity. But for many years both the B & O and Pennsy shared a healthy traffic in inbound raw materials and outbound new Baldwin locomotives at Eddystone before the company finally gave up on the business in 1956.

As short as the Crum Creek/Eddystone branch but even more important for the future was the Sparrow's Point branch, built in 1889. Really more a part of B & O's Baltimore terminal than a true Philadelphia Division operation, the 2.4-mile spur left the Philadelphia Division main line a short distance east of the Canton ferry slip and ran southeast to Colgate Creek. Here it met a private switching line, the Baltimore & Sparrow's Point, which tapped the Pennsylvania Steel Company's new integrated mill being built at Sparrow's Point, at the southeastern extremity of Baltimore Harbor. (Yes, Sparrow's Point originally had the

This tangle of trackage and freight cars was characteristic of the area near the Delaware River waterfront in South Philadelphia. The circa 1890 photograph looks along PRR's Commercial Avenue tracks toward the B & O crossing at Swanson Street. Note the unique B & O "pot" hopper cars at left center and the B & O tower at the right. B & O Historical Society Collection.

Milk shipments for one of the frequent local trains are loaded on the platform at Carpenter, Delaware about 1900. Smithsonian Institution Collection.

apostrophe in its name, now long since vanished.) Inevitably, it seemed, B & O competed with the Pennsylvania here too; the Pennsy had a similar branch to Colgate Creek to interchange with the steel company's railroad. The Sparrow's Point mill got off to a rocky start in the early 1890s but eventually became the nation's largest steel complex and a major shipbuilder — and always a heavy B & O customer. Called the Maryland Steel Company after 1891, it was absorbed by Bethlehem Steel in 1916.

With the Philadelphia Division fully operational by late 1886, B & O wasted no time in trying to make up for its lost years in the New York market. During 1887 and 1888 passenger schedules were expanded as quickly as the railroad could finish its facilities and obtain equipment. By November 1888 B & O was operating four daily through Washington-New York passenger trains each way, plus an overnight run carrying strictly express shipments.

With the same optimism that produced the multitude of pretty way station buildings, the railroad started an extensive array of local passenger services which, it hoped, would develop these underpopulated spots. In May 1887 it began running suburban trains between Philadelphia and terminals at Chester, Wilmington, and Singerly, Maryland. (Singerly, 43 miles from Philadelphia, was a tiny hamlet three miles outside Elkton, Maryland.) By November 1888 no less than 34 locals were scheduled to and from various points between Baltimore and Philadelphia: three each way between Philadelphia and Chester, nine between Philadelphia and Wilmington, two between Philadelphia and Singerly, one between Wilmington and Baltimore, and two each way all the way between Philadelphia and Baltimore. The Philadelphia-Baltimore locals took three hours and fifty minutes to struggle over the 91.6 miles between Philadelphia and the Canton ferry slip; if they made all the flag stops, they stopped and started 38 times along the way.

In fact, considering that the railroad was brand new and operating in an undeveloped territory, it was unbelievably (and perhaps unrealistically) busy. All told, the November 1888 operating timetable listed 46 passenger trains, which included the 34 locals, ten Washington-New York runs, two Philadelphia-Washington expresses, and the two package express trains. In addition, four freights were scheduled in each direction: one fast freight, one local, and two regular freights. Even the backwoods Landenberg branch was relatively active with three daily (except Sunday) passenger round trips and one local freight, actually a mixed train.

As would be expected for an interloper trying to break into long-established markets, freight traffic for the first few years was sparse. In 1888, for example, Philadelphia generated an average daily volume of about 46 inbound cars and a lean sixteen outbound. New York was even less: about fourteen daily cars in and eleven out, although at this time the Staten Island line was still uncompleted.

The number of freight yards on the Baltimore-Philadelphia line matched the volume — there was hardly any. Philadelphia's facilities consisted of a small 250-car yard along the Schuylkill River at East Side, where the Delaware River branch joined the main line, and a smaller 150-car yard at 58th Street near the temporary roundhouse. A team

As shown in these two photographs, early passenger trains were hauled by power like the 771, a new H-4 built by B & O's Mt. Clare Shop in 1887. K-3 Moguls such as the 946 handled freight. Both are seen at Philadelphia's 58th Street roundhouse in 1887. B & O Historical Society Collection, Smith photograph.

track terminal was located near the Schuylkill River at Race Street, north of the passenger station. (Team tracks were open public delivery tracks set between driveways to allow wagons ["teams"] to load and unload freight directly alongside the cars.) This and some limited trackage at Piers 62 and 63 on the Delaware River pretty much completed the B & O's Philadelphia rail freight terminals.

In March of 1889 B & O expanded its through Washington-New York passenger service to six trains each way, one of which was billed as a "Through Vestibule Limited Express Train." (Until then passenger cars commonly had open vestibules, but in 1887 Pullman began building equipment with narrow enclosed vestibules.) All trains but the overnight run carried parlor-buffet cars, and the night train handled sleepers between New York and Philadelphia, New York and Baltimore, and New York and Washington. Several of the other trains also carried through buffet-sleepers between New York and Chicago, Cincinnati, and St. Louis. The fastest 1889 New York-Washington schedule, the "vestibule limited," took five hours and forty minutes; most others consumed six hours or slightly more. Schedules were still slowed by the Baltimore harbor transfer and the awkward Reading connection at Philadelphia with its backup movement. Although the competitive Pennsylvania had neither of these problems, its New York-Washington trains used Broad Street Station in the center of Philadelphia, a stub-end terminal which required a long backup move between there and West Philadelphia.

To power its Philadelphia passenger trains — the better ones, at least — B & O assigned some of its newest locomotives. The through New York runs, usually consisting of four or five cars, were headed by a 4-4-0 with 69-inch drivers, either an H-4 class built in 1886 or 1887, an I class of 1883 to 1886, or perhaps a slightly older G class. All were essentially ordinary all-purpose engines; they were B & O's best at the time, but not specifically designed for the future competitive demands of the New York service. Two oddities, however, were a pair of center cab "camelback" 4-4-0s, Nos. 762 and 763, built by Baldwin in 1886. Class K-3 Moguls, turned out by B & O's Mt. Clare shops in 1886, handled many of the freights.

East of Philadelphia all B & O passenger and freight trains were powered by Reading and Jersey Central locomotives, and manned by Reading and CNJ crews under the 1886 three-railroad agreement. In the early years, engines typically were changed twice on the run: B & O power was exchanged for Reading at Philadelphia, and the Reading and CNJ exchanged engines at their junction at Bound Brook, New Jersey. Most commonly, the Reading assigned a 4-4-0 camelback, while at Bound Brook a handsome end-cab Jersey Central eight-wheeler took over for the last thirty miles into Jersey City.

Physically, the Philadelphia line was in more-or-less final shape by 1890. Begun late, the elaborate Philadelphia passenger station was not finished until 1888. Double tracking was completed in the Spring of 1889, along with

The John W. Garrett under full steam — although not on the B & O. This view shows the carferry after its sale to the Norfolk Southern in 1899, but its appearance and cargo are almost identical to its Baltimore Harbor career. Virginia State Library Collection, H. C. Mann photograph.

the last of the stations and freight houses. In Philadelphia, the Reading built a new connecting bridge over the Schuylkill between West Falls and East Falls. When completed in late 1889, this project finally allowed B & O trains to move directly between Park Junction and the Reading's New York branch at Wayne Junction, cutting the running time through the city in half.

But in Baltimore the unwieldy harbor carferry transfer plugged on. It did get a modest improvement in late 1886 when a larger ferry replaced the seven-year-old *Canton*. Another Harlan & Hollingsworth product, the iron-hulled *John W. Garrett*, was 351 feet long and able to carry a locomotive and nine passenger cars. Since most trains were only four or five cars, the entire train, with locomotive, usually could be run directly on and off the boat without switching. To shave more minutes off the running time, the locomotive took water from on-board tanks during its eight-minute ride across the harbor. Like its predecessor, the *Garrett* was a side-wheeler powered by a pair of single-cylinder steam engines. The *Canton* was then sold to co-owner PW & B and, ironically, sent north for Pennsylvania Railroad service in the New York Harbor.

To B & O passengers, probably the most impressive improvement in the late 1880s was at Jersey City, the gateway to New York. To cope with its rapidly growing business, which now included the B & O, its own Philadelphia services, and an expanding seashore and suburban traffic, the Jersey Central opened an entirely new terminal station in 1889 to replace its original 1864 facility. The station building itself (most of which still survives, now denuded of tracks) was a large, vaguely Romanesque iron-framed brick structure, complete with a spired clock tower. Behind it was a grand 512-foot-long iron truss trainshed spanning twelve terminal tracks and six platforms. The 75-foot-high shed had a large glass skylight running almost its entire length; combined with the station's all-electric lighting, it helped make the Jersey City terminal one of the brightest large railroad stations anywhere.

At nearly the same time, B & O finally finished its own long-delayed New York Harbor freight facilities and connections. By late 1888 all the Staten Island trackage had been completed along with the Arthur Kill bridge, but at this point the bridge's future was again thrown into doubt. The iron truss bridge had been built with a 500-foot swing center span flanked by fixed 150-foot spans at each end. The Pennsylvania and Lehigh Valley pressured the War Department to order it dismantled and replaced with a different design that would give more clearance to the water traffic.

While fighting off this latest attack, the B & O at last got around to incorporating its connecting railroad on the New Jersey side. By this time the original plan to build to the Reading at Bound Brook was changed to a short, five and one quarter-mile line directly west to meet the Jersey Central main line at Cranford, New Jersey. In October 1888 B & O created the Baltimore & New York Railway to build this stretch and, it hoped in the future, to extend southwest to Philadelphia as the B & O's own New York main line. Like the other companies created to build B & O's New York extension, the Baltimore & Philadelphia and the Schuylkill River East Side, the Baltimore & New York was purely a "paper railroad" which owned charter rights and

The Jersey Central's impressive 1889 Jersey City terminal as seen from waterside and trackside. Now restored, the station building still stands, although long since abandoned by the railroad. Exterior view from Steamship Historical Society Collection; interior from W. B. Crater Collection.

property but no equipment and was to be operated as part of the Staten Island Rapid Transit.

Work on the Arthur Kill-Cranford link started in 1889 but even this short line had its difficulties. The single-track freight route went through a developed area and crossed three other busy railroads: the Jersey Central's Perth Amboy branch, the Pennsylvania's main line near Linden, and the Lehigh Valley's new Jersey City extension. (The LV line was then under construction too.) Counting the Arthur Kill bridge and its approaches, 38 percent of the line's length consisted of bridges or timber trestles. Trestlework at the two ends of the bridge totaled 9,600 feet, 1.8 miles, which eventually was replaced by earth fill. It was not until March 1890 that trains began running to the St. George marine terminal on Staten Island.

Forever afterwards the Staten Island operation would be an odd, isolated, and very untypical B & O outpost. Separated from the nearest B & O track by 73 miles and born of a rural short line, it was left to live under its own corporate identity: the Staten Island Rapid Transit, with its own peculiar personality. The name itself had un-B & O connotations, and indeed the big railroad had little interest and enthusiasm for the short-haul rapid transit-style passenger services it had reluctantly inherited. These trains consisted of short strings of light open-platform wooden coaches similar to those used on the New York elevated lines; like the "el" trains, they were hauled by tiny 0-4-4T Forneys, 2-4-4Ts, and light 4-4-0s. B & O did little to break the pattern; during its early years under B & O control the SIRT added to its 2-4-4T fleet and also bought a small

Headed by a little Baldwin-built 2-4-4T Forney, an SIRT train poses at St. George in the 1890s. B & O Museum Collection.

collection of light 4-4-0 camelbacks plus some light-duty B & O castoffs. Except for the B & O immigrants, its equipment was always lettered "Staten Island", with no identification of its large parent.

Along with the Staten Island railway lines, B & O also inherited the ferry service between Staten Island and the lower tip of Manhattan. As part of its agreement to upgrade the local railroad services, it installed two large steel-hulled side-wheel ferries in 1888 — appropriately named the *Robert Garrett* and *Erastus Wiman*. It also rebuilt the New York ferry terminal at Whitehall Street and in 1896 commissioned the Baltimore architect E. Francis Baldwin to design an elaborate new rail-highway-ferry terminal at St. George, a structure which achieved brief and dubious fame by perishing in a spectacular fire on June 25, 1946.

In a completely separate world were the SIRT's freight operations, which essentially were run as a terminal switching railroad. Once the Baltimore & New York line was opened in early 1890, all of B & O's New York freight movements were routed through Cranford Junction, just east of the Jersey Central's Cranford station. New York-bound B & O cars were set off by Jersey Central freights, picked up by SIRT locomotives and crews, and shuttled over to St. George or Arlington yard. The SIRT also devel-

The Arthur Kill bridge's 500-foot swing section was a record length for its time.

oped a modest interchange business with the Lehigh Valley at Staten Island Junction and the PRR at Linden Junction. Although increasingly busy, the Staten Island line was far from a high-speed heavy-duty railroad, with a locomotive roster geared to its drudge work. By the early 20th Century, SIRT trains and switch runs typically were worked by 0-6-0 camelback switchers, a lone 2-8-0 camelback, and various pieces of B & O freight power — some of which were obsolescent castoffs and some of which were specifically designed for Staten Island service. (In deference to the island's growing residential development, the SIRT began to adopt "smokeless" anthracite coal in 1906. This in turn resulted in replacing virtually its entire existing steam power roster with locomotives designed to burn the fuel, primarily camelbacks with their wide Wootten-design fireboxes.) The harbor operation, however, was an odd case of corporate split personality. Essentially the Staten Island Rapid Transit ended at the water's edge. While it ran the freight trains and switched the St. George harborside terminal, all the waterborne services and New York City freight facilities were pure Baltimore & Ohio operations, using B & O equipment and people. Thus, although B & O tracks ended at Philadelphia, the company legitimately had its own New York terminals and could publish through B & O rates and bill New York freight shipments in its own name.

By the time that the last major work on B & O's Baltimore-Philadelphia line had been finished in 1889, the construction bill officially totaled $17.8 million — more than fifty percent over the original estimate. And this did not include the costs of acquiring the Staten Island lines or building the connecting line, the Arthur Kill bridge, or the freight terminals there. Neither did it include the yet-to-be-built all-rail connection through Baltimore or, for that matter, any equipment for the Philadelphia line. Ironically, back in 1881 the Pennsylvania had paid $17 million for its 92 percent interest in the Philadelphia, Wilmington & Baltimore — a price John Garrett had been unwilling to pay.

Furthermore, the addition of all the New York Extension debt to B & O's already top-heavy financial structure started the company on its way to receivership and to the end of its domination by Baltimore interests. More immediately, it led to the unceremonious ouster of the hapless Robert Garrett. By August 1887 the railroad had been forced to swallow its pride and turn for help to a financing syndicate dominated by J. P. Morgan. Morgan made short work of Garrett, insisting that he step down. The Garrett era on the B & O ended with barely a whimper upon his capitulation that October after returning from yet another trip to Europe. Unhinged by this humiliation, the Vanderbilt experience, and his other ordeals, Garrett retreated from business life entirely. Nine months later, after the accidental death of his brother, he had a complete breakdown and spent the rest of his life in seclusion. He died in 1896 at the age of 49.

Also ended was any dream the B & O still had of building its own line all the way to New York. Another Morgan ultimatum ordered the company to give up the idea and work jointly with the Reading, Jersey Central, or, as Morgan suggested, even the Pennsylvania. Although any PRR joint venture was unthinkable, it eventually did happen (though only for a short while). But certainly not in 1890, when Garrett's successors finally felt that they had a completed, first-class railroad and were fully ready for head-on competition with their arch-fiend, the Pennsylvania.

The doomed Robert Garrett already had been ousted from B & O's presidency when the SIRT received this new ferry named for him. Built in 1888, it was one of two side-wheelers bought by the B & O to upgrade the railway's New York ferry services. After the city took over the Staten Island ferry operation in 1906, it was renamed Stapleton and transferred to other services. Purslow Collection, Steamship Historical Society.

VOL. IV. DECEMBER, 1900. NO. 3.

BOOK OF THE ROYAL BLUE

TABLE OF CONTENTS.

ILLUSTRATIONS.

ALL TRAINS VIA WASHINGTON
BALTIMORE & OHIO R.R.
WITH STOP-OVER PRIVILEGE

THE POPULARITY OF THE CLUB CAR.

An early precursor of today's "in flight" magazines was "The Book of the Royal Blue," a monthly publication for passengers and the general public published between 1898 and 1911. Although created and oriented to stimulating business on the New York line, it also served as a general public relations medium for the railroad, covering points of interest throughout the system. J. J. Snyder Collection.

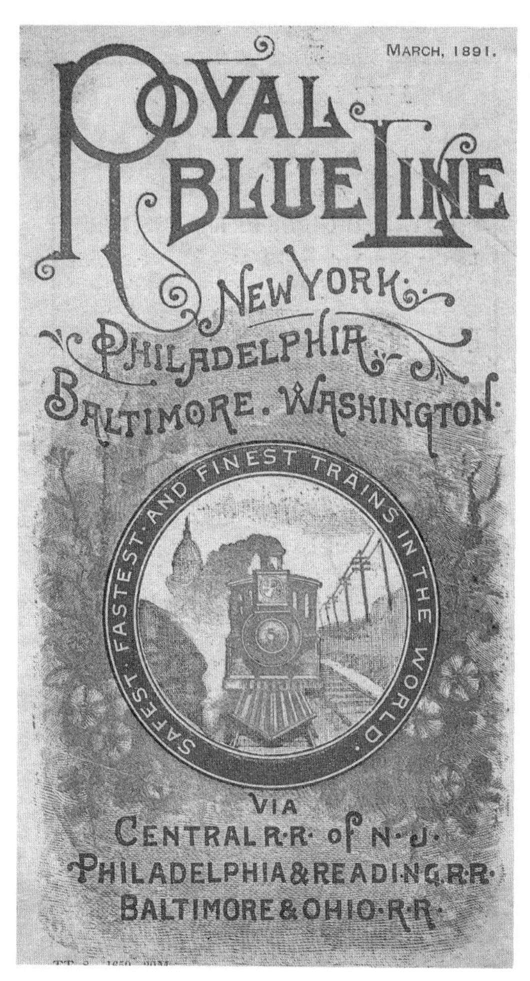

E. L. Thompson Collection, B & O Historical Society.

E. L. Thompson Collection, B & O Historical Society.

H. H. Harwood, Jr. Collection.

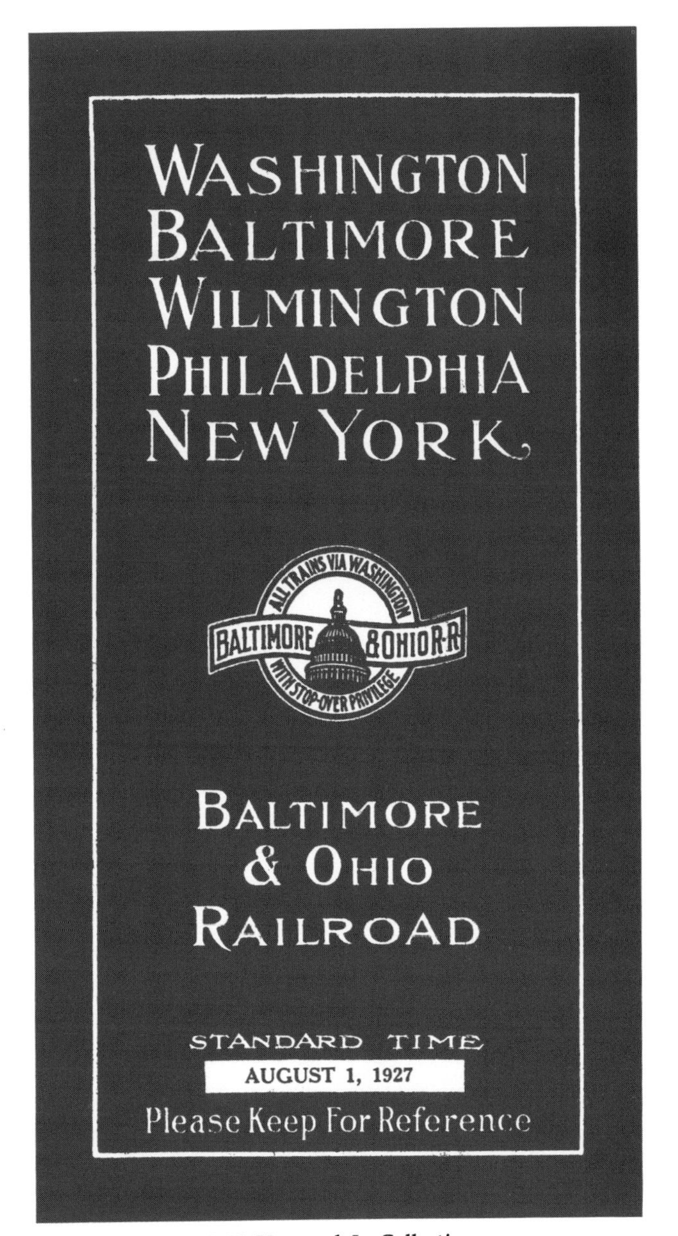

H. H. Harwood, Jr. Collection.

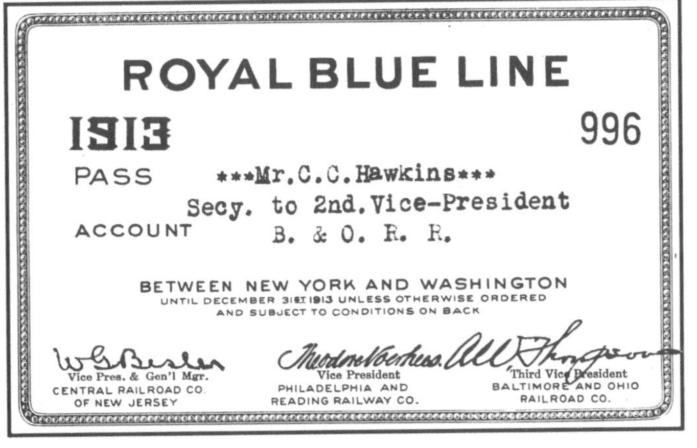

F. A. Wrabel Collection.

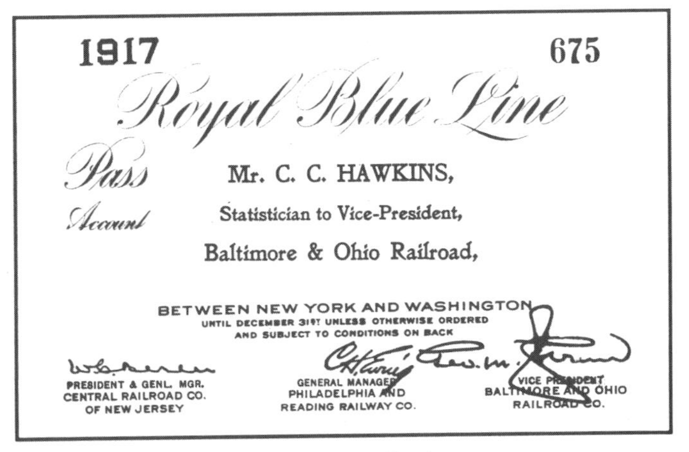

F. A. Wrabel Collection.

H. H. Harwood, Jr. Collection.

E. L. Thompson Collection, B & O Historical Society.

H. H. Harwood, Jr. Collection.

An excerpt from a 1935 folder extolling the lightweight Royal Blue. J. J. Snyder Collection.

From an elaborate brochure issued to announce the new 1936 motor coach fleet. B & O Historical Society Collection.

5 B & O MOTOR COACH *TRAINSIDE* CONNECTION ROUTES

Direct Service to and from Trainside—Leaving and Entering NEW YORK AND BROOKLYN

B&O Motor Coach Train Connection Service offers the ideal way to leave and enter New York. So easy—so convenient. You simply board a comfortable Motor Coach at any of B&O's 4 stations or 9 other stops in New York and Brooklyn (including leading hotels) and ride direct to trainside at Jersey City. When you step aboard the coach you've *made the train!*

Check your hand luggage at any B&O Motor Coach Station. It will be delivered *on the train,* to your coach seat or Pullman space. No long walks. No stairs. No traffic or baggage worries.

Entering New York you have the same convenient service. You ride directly from the trainside to any of 13 places in New York and Brooklyn—(including leading hotels). Your ticket includes this service. NO ADDITIONAL CHARGE.

MOTOR COACH STATIONS: 122 E. 42nd St. (opposite Grand Central Terminal and Commodore Hotel); Columbus Circle (Broadway and 59th St.); Rockefeller Center (49th Street and Rockefeller Plaza). Brooklyn Station (Eagle Building, Washington and Johnson Sts.)

MOTOR COACH STOPS: Hotels Lincoln, New Yorker, Taft, Victoria, Vanderbilt, McAlpin, Pennsylvania and Governor Clinton. Also Wanamakers (4th Ave. and 9th St.)

NEW YORK BAGGAGE SERVICE

All B&O Trains Handle Checked Baggage

The B&O provides a special service that relieves you of handling your baggage to New York and Brooklyn in train-connection motor coach service. Here are the details:

Inbound Service: Baggage representatives on the train will check your baggage to any B&O motor coach station or stop in New York or Brooklyn. And your baggage will be carried on the same motor coach with you. If preferred, you may check your hand luggage and baggage from starting point to any B&O motor coach station.

Attendants will gladly assist you to transfer baggage from 42nd Street Station to Grand Central Terminal, New York City.

Outbound Service: You may check your hand baggage at any B&O motor coach station in New York or Brooklyn in advance or at any time. It will be delivered to your coach seat or Pullman space on the train.

HOTEL COMMODORE
GRAND CENTRAL TERMINAL
42ND STREET
42ND STREET STATION
HOTEL VANDERBILT
33RD STREET
BROADWAY
FOURTH AVENUE
14TH STREET
9TH STREET
WANAMAKER'S
LAFAYETTE STREET
HUDSON RIVER
JERSEY CITY TERMINAL (TRAINSIDE)
LIBERTY STREET STATION
STATUE OF LIBERTY
B & O COACH ROUTE No. 1

COLUMBUS CIRCLE STATION
59TH STREET
BROADWAY
HOTEL LINCOLN
42ND STREET
HOTEL NEW YORKER
33RD STREET
EIGHTH AVENUE
14TH STREET
HUDSON RIVER
JERSEY CITY TERMINAL (TRAINSIDE)
LIBERTY STREET STATION
STATUE OF LIBERTY
B & O COACH ROUTE No. 2

10

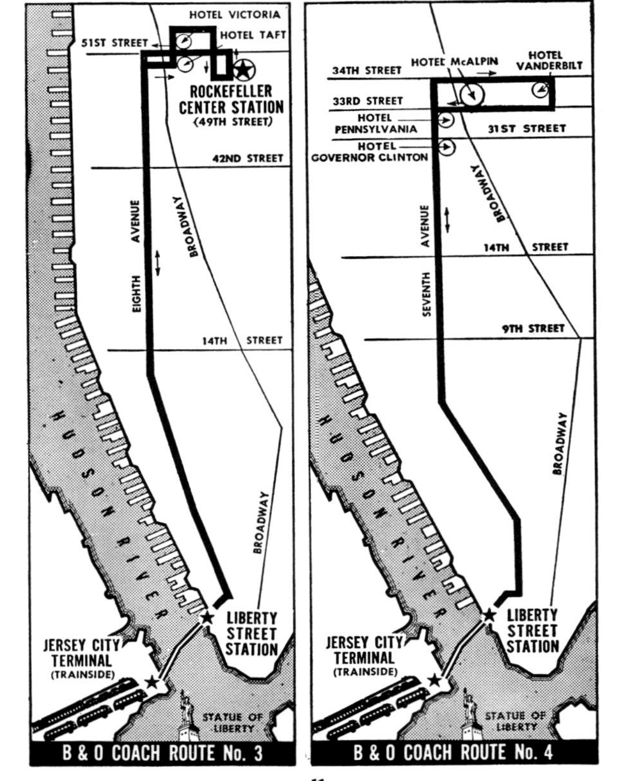

11

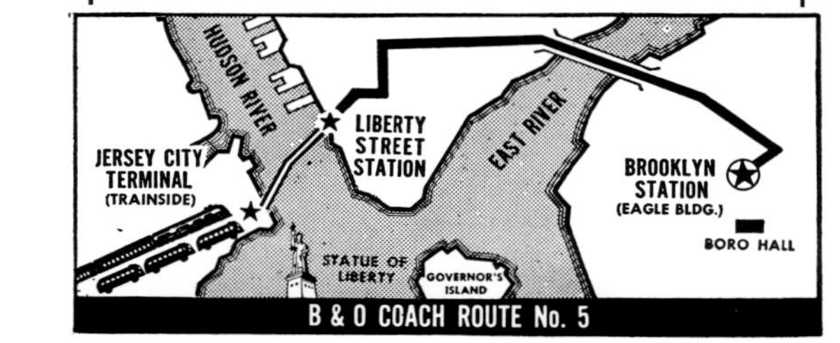

12

Typical of postwar motor coach advertising, this comes from a 1947 New York-Washington timetable. H. H. Harwood, Jr. Collection.

To expedite service, kindly write your order on check, as our waiters are not permitted to accept verbal orders. Please pay on presentation of your check.

DINNER

SOUP DU JOUR — MINTED GRAPEFRUIT SEGMENTS — CLAM BROTH

FRIED MARYLAND OYSTERS, CHILI SAUCE—2.40

BAKED SEASONAL FISH, BUTTER BASTED—2.25

ROAST MARYLAND TURKEY, DRESSING, CRANBERRY JELLY—2.50

BREADED CHOICE PORK CHOPS, SPICY TOMATO SAUCE—2.40

BROILED SELECTED SIRLOIN STEAK—3.90

CARROTS VICHY — SOUTHERN STYLE STRING BEANS

CANDIED SWEETS OR HOME FRIED POTATOES

HOT POTATO ROLLS — HOT BREAD SPECIALTY OF THE DAY

HELP YOURSELF FROM SALAD BOWL

APPLE or PUMPKIN PIE — ICE CREAM — HALF GRAPEFRUIT

BREAD CUSTARD PUDDING — MAPLE BAKED APPLE

CHOICE OF CHEESE AND CRACKERS

COFFEE — SANKA — POSTUM — TEA

MILK or BUTTERMILK

H. O. McAbee
Manager, Dining Car and
Commissary Department
Baltimore, Maryland
11-50-76

Baltimore and Ohio Dining Car Service

The end was approaching when this 1950 menu appeared, but clearly the Royal Blue remained determinedly civilized. J. J. Snyder Collection.

A shivering knot of on-leave servicemen welcomes the arrival of the eastbound Capitol Limited at Aberdeen, Maryland in December 1953. N. D. Clark photograph.

A late afternoon sun lights "A" tower and the New York-bound Royal Blue at Aberdeen in September 1953. Typical of many operating control points along the line, the tower controlled both eastbound and westbound passing sidings. N. D. Clark photograph.

The westbound Royal Blue passes Aberdeen on the first leg of its last round trip on April 26, 1958. This view is taken from the same general location as the Royal Blue shot on the previous page, but here the camera looks north from the opposite side of the track. R. L. Wilcox, F. A. Wrabel Collection.

The eastbound Capitol at Mountain Road, west of Van Bibber, Maryland, April 26, 1958. R. L. Wilcox, F. A. Wrabel Collection.

Typical of diesel-era special movements, a heavy westbound racetrack train headed by three passenger-equipped GPs loads at Aberdeen in April 1958. R. L. Wilcox, F. A. Wrabel Collection.

Although arch-competitors, B & O and Pennsy aided each other when wrecks or other problems blocked their lines. Such was the case here when three B & O F-7s took a PRR detour movement, complete with GG-1 electric, westbound, through Wilmington on October 12, 1955. R. R. Wallin Collection.

Only on the B & O: The famous "B & O Holly Tree" at Jackson, Maryland in its prime. Daylight view: B & O Historical Society Collection. Night view: F. A. Wrabel Collection.

Unremarked, and mostly unnoticed by passengers, B & O's New York Harbor "navy" handled the railroad's freight business to and from the city. Built in 1952-1953, the diesel tugs, the J. W. Phipps, Jr. and the Howard E. Simpson, were typical of harbor "motive power" during the last two decades of water operations. The ferry is the New York Central's Albany on a run between Cortlandt Street and the West Shore terminal at Weehawken, New Jersey. Both photographs courtesy of the B & O Historical Society Collection.

In a somewhat garish revival of the Royal Blue Line tradition, former Reading 4-8-4 2101 takes a "Chessie Steam Special" excursion for Philadelphia over the Susquehanna River bridge on May 14, 1977. R. H. Peterson photograph.

A string of first generation diesels led by F-7 4503 brings a westbound freight through Rosedale, Maryland in December 1973. R. H. Peterson photograp

Mixed color schemes of the Chessie System era — including a red and white Western Maryland GP-40 — move a westbound over Wilmington's solid Brandywine viaduct in May 1977. R. H. Peterson photograph.

One of the last of its kind left in the United States, the trainshed at Mt. Royal Station not only still survives, but still spans an active main line. GP-40-2 No. 4107 is westbound in October 1979. F. A. Wrabel photograph.

The "new" Royal Blue Line meets the old: A Philadelphia-bound TOFC train powered by a pair of wide-open GP-40-2s roars past the 1886 station at Aberdeen, Maryland. The date was 1986, an even 100 years after the station was built. R. H. Peterson photograph.

Necessity and good politics have made the old Royal Blue Line ecumenical, accommodating even the ancient enemy. Conrail, the Pennsylvania's lineal descendent, now regularly uses the route to avoid the one-time PRR main line, which in turn is Amtrak's high-speed domain. Here Conrail's train SEPY for Potomac Yard crosses the hallowed Thomas Viaduct at Elkridge, Maryland in November 1982. H. H. Harwood, Jr. photograph.

SPEED, COLOR AND CLASS: THE ROYAL BLUE LINE

The Early 1890s

There was no honeymoon time for B & O's New York Extension. The project was an enormously expensive gamble for a financially weak company; furthermore it faced a plainly powerful and well-established competitor. The first three years of the line's life necessarily was spent in finishing facilities, organizing operations, and designing and ordering equipment appropriate for the services. But in 1890 everything rapidly came together.

It was a dazzling year indeed: a new image, new and extra-luxurious passenger cars, high-speed locomotives to haul them, the new Staten Island and Manhattan freight terminals, and, at last, the start of work on the Baltimore Harbor bypass.

Most visible and memorable were the passenger trains and the image that went with them. In a single move they became the finest on the B & O system and were outstanding by the standards of any railroad. Under B & O's direction, the three New York-Washington route partners , B & O, Reading, and Jersey Central, jointly ordered a fleet of 28 deluxe passenger cars from the Pullman Palace Car Company. Car ownership was split roughly according to the terms of the 1886 traffic agreement: seventeen to the B & O, nine to the Reading, and six to CNJ. In all, there were nineteen day coaches, six coach-baggage combines, and seven baggage cars. Together with B & O diners and Pullman parlor cars and sleepers, there was enough equipment for ten re-equipped train sets.

The coaches set a new level of luxury. Large for their time (67.5 feet end to end, with a sixty-foot car body), they seated 56 passengers and included a separate nine-seat smoking room and unusually commodious toilet facilities. Interiors were paneled in light mahogany; seats were up-

The apex of the late Victorian "day coach." The Maryland state seal at the car's center indicates B & O ownership. Smithsonian Institution Collection, Pullman photograph.

Inside a Reading-owned Royal Blue Line coach. Note the carpeting, overhead racks, and, of course, the Pintsch gas lamp fixtures and ceiling decoration. Smithsonian Institution Collection, Pullman photograph.

holstered in old-gold plush. Incorporating the newest in passenger car technology, all of them (even the baggage cars) incorporated fully enclosed vestibules, using the Pullman Company's pioneering narrow vestibule design, and included steam heating and Pintsch gas lighting.

If the interiors were first class, the exteriors were in a class by themselves. Basically one of the handsomest designs anywhere, the big coaches and combines had majestic paired windows with ornately-etched, frosted glass fixed transom windows; platform railings were iron scrollwork.

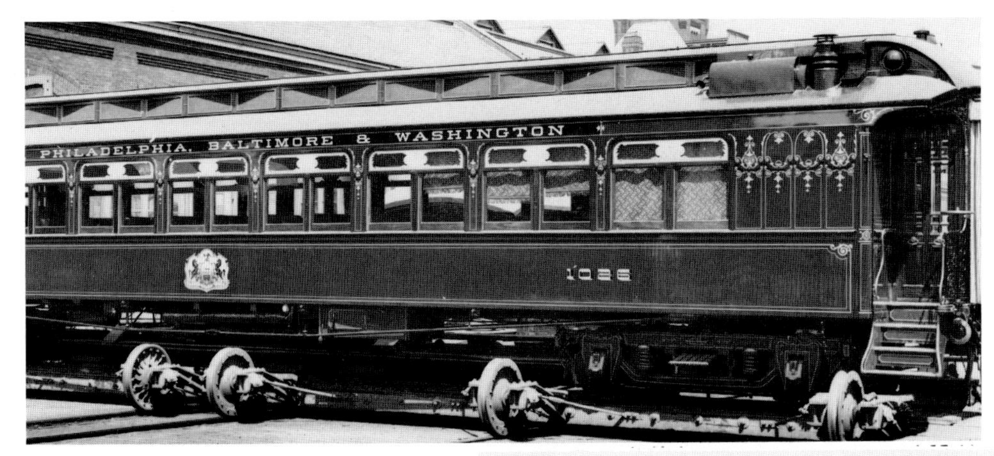

This partial view of Reading-owned Royal Blue Line coach 1026, posed on Pullman's transfer table, graphically emphasizes its elaborately fussy exterior trim and painted Pennsylvania state seal. Smithsonian Institution Collection, Pullman photograph.

Almost as ornate were the baggage cars, as amply demonstrated by this Jersey Central example. Smithsonian Institution Collection, Pullman photograph.

Then there was the color and lettering: the symbols of the service and the image that would last for almost three decades. All cars were finished in a deep "royal blue," as it was called, with delicate gold striping and painted scrollwork. No railroad name appeared on the car; instead, the entire fleet carried the legend "NEW YORK, PHILADELPHIA, BALTIMORE & WASHINGTON" on the letterboards, and at each car's center was a full-color state seal symbolizing its owner: Maryland for the B & O, Pennsylvania for the Reading, and New Jersey for the Jersey Central.[1]

From the color came the name of the service and the railroad route itself: The Royal Blue Line. The newly-equipped trains were generically advertised as "Royal Blue Line Service" or "Royal Blue" trains. (At this time there was no single specific train named *The Royal Blue*, and in fact there was none until 1935 when the name was revived after almost two decades of disuse.) For the next quarter century

[1] The "Royal Blue Line" coach now displayed at the B & O Railroad Museum in Baltimore is not one of these cars. Although an authentic narrow vestibule coach of the 1890 era, it is shorter and considerably less opulent. This car originally was built for some other service and was "re-created" as a Royal Blue Line coach for the B & O's Centennial celebration in 1927. It has been kept in its spurious livery ever since; but legitimate or not, it is a rare survivor of the short-lived narrow vestibule design.

Royal Blue Line power: Gleaming new M-1 class No. 855 at the Baldwin Works in 1892. These worked on the B & O section of the run, between Washington and Philadelphia. Baldwin Locomotive Works photograph.

Sister 857 some years later, after the class had been demoted to lesser duties. H. H. Harwood, Jr. Collection.

the Royal Blue Line was the most prominent and proudest name in B & O's lexicon; the services were given the leading position in B & O timetables and in its *Official Guide* listings, usually along with some special advertising. "Royal Blue" quickly caught everyone's imagination and became part of the railroad travel vocabulary.

The specific shade of blue, incidentally, was carefully chosen with great effort by B & O's then-president Charles F. Mayer. George Shriver, his secretary (and later B & O's senior vice president) recalled later that Mayer "had been much impressed with a certain blue, but could not have it reproduced to his satisfaction. He kept after it, however, until he finally secured from Europe a bit of old velvet of a shade known as 'Royal Saxony Blue' (the particular shade he desired) so there was reproduced that deep, rich blue, and the line became known as the 'Royal Blue Line.'" Shriver then went on to say that the color really was not entirely successful, since in time the blue faded irregularly into varying shades.

SAFEST
FASTEST
FINEST

TRAINS IN THE WORLD

ARE THE

ALL TRAINS RUN VIA WASHINGTON

B&O

Royal Blue Line Trains

BETWEEN

NEW YORK, PHILADELPHIA,
BALTIMORE, WASHINGTON,

RUNNING VIA

Baltimore and Ohio Railroad.

All trains are **Vestibuled** from end to end, **Heated** by **Steam**, **Lighted** by **Pintsch Gas**, **Protected** by **Pullman's Anti-Telescoping Device**, and operated under **Perfected Block Signal System.**

The Baltimore & Ohio Railroad

Maintains a complete Service of Vestibuled Express Trains between

NEW YORK,
CINCINNATI,
ST. LOUIS,
AND CHICAGO,

EQUIPPED WITH

Pullman Palace Sleeping Cars,

RUNNING THROUGH WITHOUT CHANGE.

ALL B. & O. TRAINS

BETWEEN THE

East AND West run via Washington.

Through the Vestibule

PRINCIPAL OFFICES:

211 Washington St., Boston, Mass.
415 Broadway, New York.
N. E. Cor. 9th and Chestnut Sts., Phila., Pa.
Cor. Baltimore and Calvert Sts., Baltimore, Md.
707 15th St., N.W., cor. N.Y. Ave., Washington, D.C.
Cor. Wood St. and Fifth Ave., Pittsburg, Pa.
Cor. Fourth and Vine Sts., Cincinnati, O.
193 Clark St., Chicago, Ill.
105 North Broadway, St. Louis, Mo.

R. B. CAMPBELL, General Manager. } Baltimore, Md. { CHAS. O. SCULL, General Passenger Agent.

To power its Royal Blue Line trains, B & O had to break out of its old motive power mold. Traditionally, its steam locomotives had been designed to cope with sharp curves, steep grades, and sometimes bridge limitations; they tended to be heavy-duty slow workhorses. With its New York services B & O was operating a different railroad in a different competitive world; it needed engines specifically designed for fast running under less limiting operating conditions. Beginning in 1890 the Baldwin Locomotive Works turned out a group of ten heavy 4-4-0s with 78-inch drivers, the largest the B & O had ever used. Produced through 1893, these impressive class M-1 engines sired a new breed of fleet, handsome high-speed locomotives for the B & O. Both Baldwin and the railroad were so proud of the design that they exhibited examples at the 1893 Columbian Exposition in Chicago. Baldwin showed off No. 858, the last "conventional" M-1 built, and B & O's exhibit featured a Vauclain compound version, No. 859, named *Director General.*

The Royal Blue Line trains were officially inaugurated on July 31, 1890 with six schedules each way: five day trains and an overnight run. The premier train, a midday schedule, covered the 226 miles (including the Jersey City-Manhattan ferry) in five hours flat. (Actually, this tight schedule had been started two months earlier to meet the time of the Pennsylvania's *Congressional Limited.*) Other Royal Blue schedules were more relaxed, taking anywhere from five hours forty minutes to six hours. As before, through sleepers and sleeper-buffet cars were carried between New York and Baltimore, Washington, Chicago, Cincinnati, and St. Louis. Perhaps not exaggerating, B & O

M-1 4-4-0 No. 853 pilots a short train, probably a local or Philadelphia-Washington run, past the picturesque station at Twin Oaks, Pennsylvania about 1900. John C. Hayman Collection.

advertised its Royal Blue trains as "The Fastest, Finest and Safest Trains in the World."

In addition to these through trains and the bustling but ordinary local services, B & O scheduled a Philadelphia-Washington semi-express run each way, primarily geared to business and political "commuters" from Philadelphia. The Washington-bound leg of this run started in the morning from the Reading's Wayne Junction station in northwest Philadelphia rather than the B & O's own downtown Chestnut Street terminal. The idea apparently was to catch inbound early morning commuters from the suburbs and also passengers from the Reading's overnight trains from upper New York state and various central Pennsylvania points. With occasional variations, this Wayne Junction-Philadelphia-Washington service survived until 1956.

All three Royal Blue Line partners worked to whittle down running times and raise capacity. In 1890 the Reading installed track water pans at Yardley, Pennsylvania, allowing steam locomotives to take water on the fly; the Jersey Central followed suit in 1891 with a set of track pans at Green Brook, New Jersey, west of Dunellen. (Later the Reading relocated its pans two miles west to Roelofs, Pennsylvania.) Also in 1891 the Reading and CNJ stopped changing engines at Bound Brook and pooled their power between Jersey City and Philadelphia.

The two railroads also followed, and in some cases surpassed, the B & O with high-speed steam power. Between 1892 and 1893 the Jersey Central bought eleven end-cab 4-4-0s with 79-inch drivers, looking quite similar to B & O's M-1s. The Reading, always a nonconformist

This tiny, charming station served Elk Mills in Cecil County, Maryland. B & O Museum Collection, B & O photograph.

railroad, simultaneously installed a fleet of eleven Baldwin-built 2-4-2 camelbacks with 78-inch drivers. Called "Columbians" in honor of the 1893 Exposition, these fast but slippery engines were among the few of this wheel arrangement ever built for American main line service. Stranger, and slipperier still, were two 4-2-2 "Bicycle"-type camelbacks built in 1895 and 1896. Assigned briefly to some "Blue Line" trains, these Vauclain compounds rode on huge 84-1/4-inch single drivers; they could turn in respectable speeds with five cars or less, but struggled with anything longer.

Concurrent with its new Jersey City terminal project in 1889, the Jersey Central also completed four-tracking its main line between Jersey City and Bound Brook. And between 1891 and 1892 it installed automatic electro-pneumatic semaphore signals over the same stretch of straight, fast track — now busy with the Philadelphia and Washington passenger expresses, suburban trains, merchandise freights, and a heavy anthracite coal traffic into New York.

On its Philadelphia Division, B & O also built track water pans in 1893 at Swan Creek, Maryland (a convenient stream between Aberdeen and Havre de Grace) and at Stanton, Delaware, giving a grand total of four sets of pans along the full Royal Blue Line route. By this time too, B & O's portion of the line had been fully fitted with block signals and rebuilt with 85-pound rail.

Passenger train operations were tightly managed and were every bit as fast as advertised. One observer in 1892 rode the five-hour train and noted: "If the train is over four minutes late, the engine driver and fireman have to report and give an explanation." Returning on a six-car regular train he timed some 65 mph running on the B & O section, and up to 68 mph on the Reading and CNJ lines.

Some trains did far better than that, especially on the Reading-Jersey Central section, where the track was straighter. Said the trade magazine *Railway Age* in 1895: "The stretch of 90.2 miles of Philadelphia & Reading and Central Railroad of New Jersey track between Philadelphia and Jersey City forms the race course over which has been attained, for short distances, the marvelous rate of 97.3 miles an hour, with numerous examples of 85 mph and more in 1890, 1891, 1892 and 1893. Indeed a locomotive builder says 'It is a very common occurrence for engines handling Blue Line trains to reach 90 mph.' "

Track water pans like this at Swan Creek, Maryland helped speed Royal Blue Line services until the end of steam in the 1950s. The experimental 4-6-4 Lord Baltimore and lightweight Royal Blue streamliner demonstrate the operation in 1935. R. W. Breiner Collection, B & O photograph.

A Royal Blue Line train and its M-1 locomotive ride across Baltimore Harbor aboard the John W. Garrett about 1890. Note the hose on the deck at the right, used to replenish the engine's water supply during the trip. The abundance of Sunday-best-dressed passengers suggests an inaugural trip of some kind. DeGolyer Library Collection.

Somewhat strangely the New York-Washington running times backslid slightly in 1893. In January of that year the Reading opened its grandiose new Reading Terminal at 12th and Market Streets in Philadelphia. Starting in May, all B & O trains except the five-hour schedule were routed through Reading Terminal in addition to B & O's own 24th and Chestnut Streets station. To accomplish this, they had to be backed over two miles across the Reading's City branch (which followed Pennsylvania Avenue and Callowhill Street) between Park Junction and Reading Terminal. B & O locomotives took eastbound trains as far as Park Junction where a Reading switcher picked them up for the move over to Reading Terminal. There, a Reading or Jersey Central engine would be coupled on for the run to Jersey City. The reason for this detour is undocumented, but possibly it was to enable B & O to compete better with the Pennsy's more centrally located Broad Street Station. In any event, the operation lasted only until early 1895 when it was slowly phased out; the last such service ended in late 1896 when the overnight train was taken out of Reading Terminal.

Another oddity of the early 1890s was an even shorter-lived through Pullman service between New York and New Orleans. Although B & O never had been able to operate into the South through Washington, it did make an attempt through another gateway. In 1893 it advertised a New York-New Orleans sleeper-buffet car, which traveled over the Royal Blue Line from New York to Washington, then continued west on B & O's main line to Shenandoah Junction, West Virginia, west of Harper's Ferry. At Shenandoah Junction the car was turned over to the Norfolk & Western Railway, continuing south over the N & W, the East Tennessee, Virginia & Georgia, and the Queen & Crescent Route. The trip took close to 45 hours. As usual, the Pennsylvania offered a similar service over much the same route via Harrisburg, Pennsylvania. After the demise of this run, B & O again abandoned the South until 1916, when another New York-New Orleans sleeper was inaugurated — this one routed through Cincinnati, Louisville, and the Illinois Central Railroad. That too died soon, a victim in 1918 of World War I.

Less advertised and certainly less noticed were the New York freight operations which, in truth, were the Royal Blue Line's major reason for existence. But in their way, these were even more fascinating, largely because the railroad became waterborne. As mentioned earlier, the key year of 1890 also marked the opening of B & O's own New York Harbor terminal. On March 16, 1890, three and one half months before Royal Blue Line passenger services started, B & O freight was diverted from Jersey City to Staten Island, using the new line from the Jersey Central at Cranford. A harbor transfer terminal had been built at St. George next to the SIRT's temporary passenger train and ferry terminal at the northeast corner of the island.

New York traditionally was unique among American cities in the enormous amount of freight which had to be floated from one part of the port to another. Steamship piers, warehouses, local industries, and all manner of other freight customers were scattered all around the fragmented harbor, usually separated from the equally helter-skelter railroad terminals by some waterway. Manhattan Island itself was reached directly by only one railroad, the New

These two photographs show the early B & O New York waterfront freight terminals. ***Above:*** *B & O's only railroad tracks on Manhattan were at the West 26th Street yard and carfloat terminal, about 1920.* ***Below:*** *The Pier 22, North River, handled less-than-carload freight loaded and unloaded from carfloats secured on the river side of the building. Both photographs from B & O Museum Collection.*

York Central; at that time there was no all-rail route to Brooklyn whatsoever. So from their earliest days, all of New York's railroads developed marine operations as extensions of their tracks. It was a costly curse but the necessary price of doing business in the country's largest city. By the 1890s flocks of smoking, whistling railroad tugs crisscrossed the harbor in bewildering patterns, nosing their barges and carfloats carefully among the ferries, ocean freighters, passenger liners, and excursion boats. It was a special and highly skilled brand of railroading, especially at night and in snow, rain, and fog.

Although somewhat smaller than many, B & O's base at St. George was typical of the area's railroad harbor terminals. Essentially, it handled three different types of rail-water freight transfer jobs: lighterage (or "break bulk") shipments, bulk transfer, and carfloating.

Lighters were covered barges, used mostly to pick up or deliver freight directly at a ship's side or at some local waterfront industry. Lighterage freight had to be laboriously transferred between railroad cars and the lighters by armies of railroad workers, sometimes aided by derricks or cranes. Bulk commodities — coal, in the B & O's case — were much easier; they were simply dumped from hopper cars into ships or open barges (called scows), using a trestle or a mechanical car dumper. Carfloating, of course, meant carrying the railroad cars with their loads intact across the water on specially-built barges equipped with rails.

B & O and the other railroads used their carfloats mostly for two types of business — pier deliveries and interchange. The piers, in effect, were waterside rail freight stations; the carfloats were tied up alongside the covered piers and the cars were loaded or unloaded as they sat on the water. Most of these piers served as freight consolidation and distribution centers, unloading the cars for drayage delivery around the city and receiving shipments brought in by the wagons. Some specialized in perishables, others in general less-than-carload freight. Again, hordes of laborers were kept constantly busy.

By the turn of the century, B & O operated two pier terminals on the Hudson River in Manhattan — Pier 7 at Rector Street and Pier 22 at Harrison Street. More were added in later years, both on the Hudson and East Rivers and, for a while, at Wallabout in Brooklyn.

In addition to its pier stations, B & O established a small "inland" railroad yard and freight handling complex adjacent to the Hudson at West 26th Street between 11th and 13th Avenues, close to the city's retail district. Established about 1890, the 26th Street yard was B & O's only true railroad facility on Manhattan, and was switched by a B & O (not SIRT) locomotive and crew. Like the pier stations, it was connected to St. George by carfloat.

The New York-area railroads also extensively used carfloats to interchange cars with one another, especially traffic to and from New England (via the New Haven Railroad's Bronx carfloat terminal) and for the Long Island Rail Road at Long Island City in Queens. But the B & O did not. It was one more symptom of the company's odd "there but not there" status in New York. While B & O originated and terminated freight at its own terminals around the city, New York did not exist as a B & O interline junction point. For freight movements which passed through the city, such as a shipment from Baltimore to Boston, the B & O ended at Philadelphia; east of there the shipment was a pure Reading-Jersey Central movement, and the CNJ interchanged it with the New Haven in New York. Aside from some local terminal switching railroads, B & O's only railroad-to-railroad carfloating in New York harbor consisted of business to and from local points on Staten Island which was interchanged with the New York Central (at both Weehawken, New Jersey and 72nd Street in Manhattan), the Lackawanna, and the Erie.

This crude 1891 facility at St. George transferred coal from rail cars to barges for delivery around New York Harbor. The view dates to about 1911. B & O Historical Society Collection, B & O photograph.

In addition to general merchandise business, B & O determined to make New York an outlet for bituminous coal from its Maryland and West Virginia mines.

In 1891 the first coal transfer dock was built at St. George, a primitive affair consisting of a single timber trestle which unloaded into barges. Another wooden trestle followed a year later and the business continued to expand to become, for some years, B & O's largest single inbound commodity at the port. In addition to industries and electric utilities, much B & O coal was used for vessel fuel in an era when virtually all ships of any type also were steam powered and coal fired.

Back on B & O's Philadelphia Division, another short freight branch appeared in the early 1890s — the last one to be built on the line. In 1892 a spur was built between the main line at Childs, Maryland (northwest of Elkton) and Providence Mill, four and one half miles to the north. Known as the Lancaster, Cecil & Southern branch (invariably shortened to the "LC & S branch"), it followed a wild, wooded gorge cut by Little Elk Creek and served several small paper mills.

The branch's name implied something more ambitious, and that was certainly so. Originally it was not even intended as a B & O line, but rather, it was planned as part of a chain of railroads which would carry anthracite coal from the Reading's eastern Pennsylvania mines to Baltimore. The coal trains were to rumble south through Reading, Lancaster, and Quarryville, Pennsylvania and then to Childs, where B & O would haul them into Baltimore.

This odd route was to be patched together from a combination of some Reading-controlled branches, new construction, and, strangest of all, a rural narrow gauge line. In 1890 the ever-expanding Reading had leased the Reading & Columbia, giving it a line between Reading and Lancaster, Pennsylvania. Along with the Reading & Columbia came another line called the Lancaster & Reading Narrow Gauge Railroad, which continued southeast from Lancaster to Quarryville, Pennsylvania. Despite its title, the Lancaster & Reading Narrow Gauge was standard gauge and, in effect, gave the Reading a through route from its main lines at Reading to Quarryville, which lay about thirty miles northwest of B & O's Philadelphia main line.

A local and its white-capped crew on the one-time Wilmington & Western at Landenberg, Pennsylvania, taken about 1900. A typical branch line hand-me-down, G-3 class No. 638 dates to 1865. B & O Historical Society Collection.

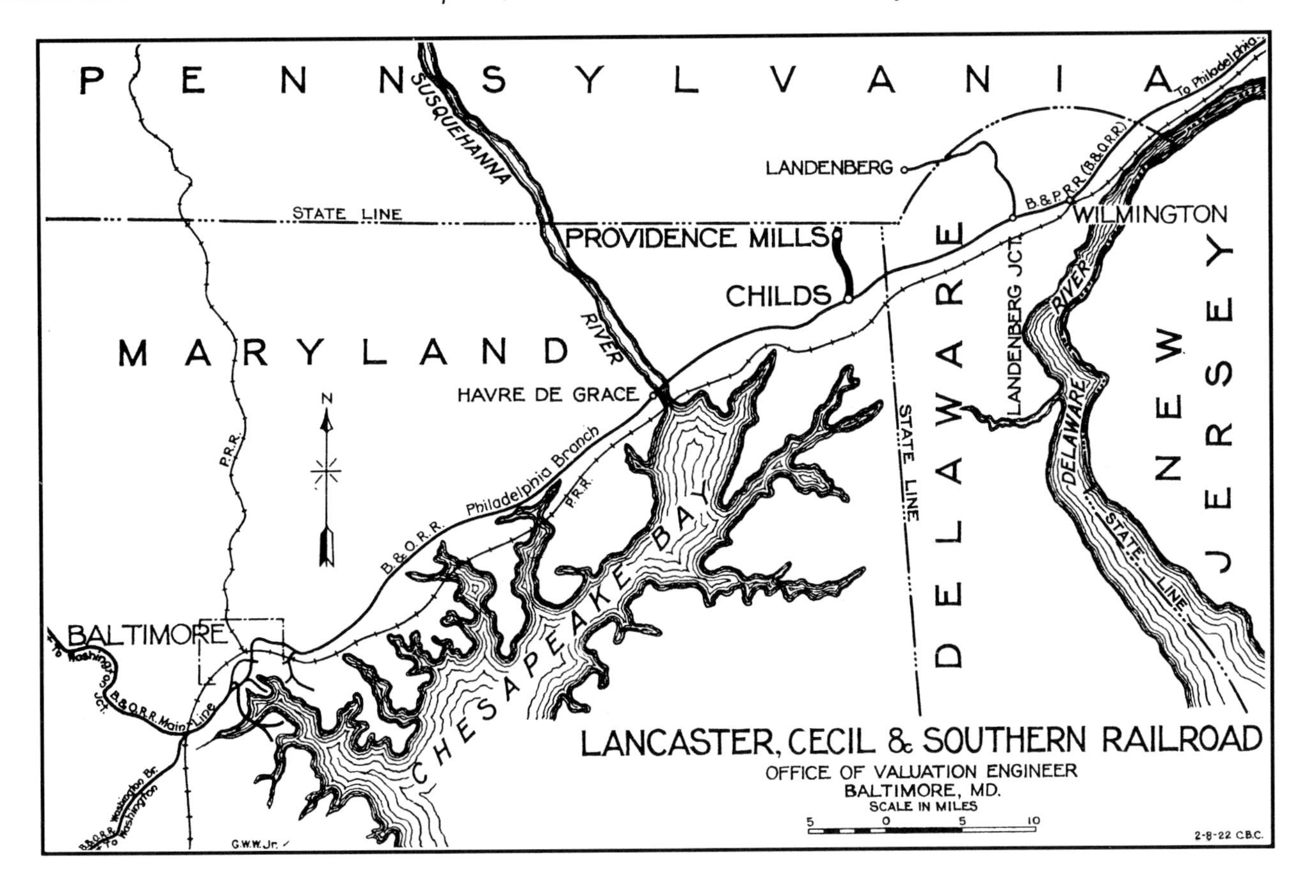

Quarryville itself was nowhere, a small country town with no other rail connections. But midway between Quarryville and the B & O was the struggling Lancaster, Oxford & Southern, a genuine three-foot-gauge railroad eking out a precarious life on a starvation diet in the hilly backwoods country between Oxford, Pennsylvania and the Susquehanna River at Peach Bottom. The LO & S's predecessor, the Peach Bottom Railroad, had gone bankrupt in 1890 and was sold to a group of local businessmen who reputedly were backed by the B & O. A plan quickly evolved to extend the LO & S in two directions to form the missing link between the Reading at Quarryville and the B & O. Needed would be an eight-mile branch northwest from the LO & S line to Quarryville, and about twelve miles of new railroad from the LO & S's eastern terminal at Oxford to the B & O at Childs. The existing narrow gauge center section of the route would be rebuilt with dual gauge track.

The Oxford-Childs extension was incorporated in February 1892 as the Lancaster, Cecil & Southern. Technically the LC & S was owned by the owners of the Lancaster, Oxford & Southern, but actually it was supported by cash advances from the B & O. Construction of its first segment, the four and one half miles from Childs to Providence Mill, was started that spring and the line opened in May 1893.

And that, of course, was as far as the LC & S ever got. The B & O, by then sinking into serious financial trouble and trying to complete its costly Baltimore Belt Line, could not find enough loan money for its ambitious anthracite adventure. In the meantime too, the project had been blighted by another development: in February 1893 the Reading went into receivership (its third) and defaulted on the bonds of the unproductive Lancaster & Reading Narrow Gauge branch. The Lancaster & Reading was sold a year later, removing it as an affiliated link in B & O's planned anthracite route. Although not worth much on its own, it was soon absorbed by the Pennsylvania to prevent any future B & O poaching.

Though operated by the B & O, the aborted LC & S remained a separate company for several years afterwards. Finally in 1899 B & O took over ownership in exchange for its cash advances, and afterwards the short branch lived off the ever-dwindling paper mill traffic. Before its final abandonment in 1972, it had been reduced to slightly more than a mile and was used mostly to store cars.

But if the LC & S led the B & O nowhere, a somewhat similar (and even more abortive) project helped it to break its Baltimore impasse. Originally, the company had hoped to connect its Camden station with the Philadelphia Division by building an elevated line around the north side of the harbor, following Pratt Street and other streets near the shore line. This solution had the virtues of being gradeless and open, but it was also expensive and controversial. The route necessarily had to pass through built-up residential areas near Fells Point and Highlandtown; property owner opposition helped shortstop it before work was started.

An alternative appeared in 1888. The Maryland Central Railroad was a short line and, coincidentally, also narrow gauge, which twisted and climbed its way north from Baltimore to Delta, Pennsylvania via Belair, Maryland. At Delta it met another narrow gauge line, the York & Peach Bottom Railway which, by even worse contortions, extended northwest to York, Pennsylvania. (The two lines later were standard gauged and in 1901 were merged as the Maryland & Pennsylvania, the "Ma & Pa.") Built between 1881 and 1883, the Maryland Central was an underfinanced latecomer to Baltimore which ended at a remote terminal in the Jones Falls valley at North Avenue and Oak Street (later Howard Street), on the city's northern fringe. From there its line ran through delightfully pretty rural country, minimally populated and innocent of any major traffic potential.

Ambitious if nothing else, the Maryland Central hatched a scheme to turn itself into a Baltimore outlet for anthracite coal by extending north to connect with two major Pennsylvania coal haulers, the Reading and the Lehigh Valley. To do so, the MC would build from its existing line at Belair north to Conowingo, cross the Susquehanna, and continue north to the Reading's main line at Birdsboro, Pennsylvania, then swing northeast to join the Lehigh Valley at Bethlehem. Construction north from Belair toward the river began in 1889.

With its Baltimore terminal far uptown and any further extension south down the Jones Falls valley blocked by the Pennsylvania's main line and its branch to Calvert station, the Maryland Central needed a way to get to Baltimore Harbor and the downtown area for coal delivery terminals. One idea was to build a long branch east to the B & O's Philadelphia Division at Herring Run, northeast of the city. Although the early plans were vague, the two railroads decided to join forces to solve their different Baltimore terminal problems. In December 1888 they jointly incorporated the Baltimore Belt Railroad, which was to form the Maryland Central's Baltimore entry and the B & O's connection between its Camden station and the Philadelphia line. The Maryland Central's North Avenue terminal station lay about two miles directly north of Camden along the line of Howard Street; a rail route could tunnel under Howard Street, connect with the MC, then turn east to meet the Philadelphia Division on the far east side of the city.

The Maryland Central's ambitions absurdly outreached its anemic financial resources; work on its Belair-Conowingo line quickly petered out with only some grading accomplished and it soon withdrew from the Baltimore Belt. By 1890 B & O had assumed full control of the project, although the Baltimore Belt Railroad remained a separate company with separate financing. With B & O's own financial situation clearly deteriorating, it was in a poor position to raise the needed money — which would be considerable — with its own credit.

As finally designed, the Belt Line was to begin at Camden, go directly north under Howard Street, cross over the Maryland Central at Jones Falls near North Avenue (there was still lingering hope of a connection), continue north to what is now 26th Street and Huntingdon Avenue, then head eastward across Baltimore's far north side. It would join the Philadelphia line at Bay View, coincidentally next to the old PRR-PW & B junction at the city's northeast corner. In all, the Belt would be only 7.3 miles long, but they were seven very difficult and costly miles. Yet expense was the lesser problem. To build and operate the Belt, B & O would need to call back the creativity and daring of its earliest days. What it planned to do required technology which barely existed, and not at all in the form B & O needed.

PIONEERING AGAIN: BUILDING THE BELT LINE

1890-1896

The Baltimore Belt franchises were cleared in 1890 and design work proceeded with urgency. The project's 7.3-mile length was an almost continuous succession of engineering challenges; included were the 1.4-mile-long Howard Street tunnel, six short tunnels in North Baltimore, a viaduct over Jones Falls valley, several other overpasses, and considerable cut-and-fill work. Apparently lacking anyone with the needed skills in his own company, B & O president Charles Mayer raided the rival Pennsylvania and hired Samuel Rea as the Belt Line's chief engineer. Rea, then only 34, was an exceptionally capable engineer who had risen quickly in the Pennsy's ranks, but was then somewhat frustrated in his position of Assistant to the Vice President.

The Belt's cost was estimated at $6 million — $5 million for actual construction plus an extra million for contingencies and later improvements. Some creativity had to be used in raising the money, since B & O's debt already was excessive and the company was in a poor position to guarantee any bonds issued by the Belt Railroad. The problem was solved by negotiating a mortgage in the Belt Line's name, which was tied to a long-term rental contract with the B & O. In this way, Belt Line bondholders would continue to receive their income if the B & O went into receivership, something that B & O executives were beginning to fear.

Construction contracts were let on September 12, 1890 and, except for the tunnel, work began soon afterward. Not surprisingly, it turned out to be grueling and long, lasting over five years. Topography and geology created considerable problems in design and management of the project, which were aggravated by the urban and suburban environment through which much of the line was to pass. The city's downtown section where B & O's Camden station and its other yards and terminals were located lay in a bowl surrounded by modest hills. The northward route of the Belt Line underneath Howard Street required a steady climb of 0.8 percent as far as the tunnel's north portal; from there the grade steepened to over one percent as the line worked its way across the Jones Falls valley and up the other side, hitting a stiff 1.55 percent before reaching its summit at Huntingdon Avenue. At that point the route turned abruptly east, running parallel to the present 26th Street through the soon-to-be-developed residential community of Peabody Heights (now called Charles Village) and the older settlement of Waverly at York Road. Through these areas the railroad had to be as inconspicuous and unobtrusive as possible, and the entire route was to be free of any grade crossings.

The Howard Street tunnel posed particularly nasty problems, not only to build but also to operate afterwards. The street was a main north-south thoroughfare and was built up along most of its length. Lining it were residences, several fashionable department stores, many commercial establishments, the Academy of Music, City College, and, adjacent, Johns Hopkins University. In addition to its heavy carriage and wagon traffic, Howard Street carried a horse car line (which would be rebuilt and electrified during the course of the tunnel construction) and for three blocks a cable-powered street railway which was just then being built. Underneath the street was mostly sand and clay, including some underground streams and quicksand, which had to be dealt with without disturbing the surface traffic or buildings. And, at 7,340 feet, the Howard Street tunnel would be the country's longest soft-earth tunnel and the longest on B & O of any type.

All of that was merely the construction challenge. More worrisome was how the tunnel could be operated once it was finished. Its length, combined with the relentless 0.8 percent northward grade over its entire distance, promised problems with steam locomotive smoke and toxic gases. The Pennsylvania Railroad's Baltimore tunnels were an all-too-immediate case in point, which neither the city nor the B & O wanted to see repeated. Twenty years earlier the Pennsy had paid a stiff price for its bold invasion of the B & O's city. To cope with some of the same conditions which the B & O now faced, it had been forced to build a series of four tunnels totaling 11,074 feet in length across the city's north side. One, the Union tunnel east of the PRR passenger station, had a 1.2 percent eastbound grade. But much worse were the three west of the station, collectively called the "B & P tunnels." Named for their nominal owner, the Baltimore & Potomac Railroad, these actually consisted of a single 7,669-foot bore punctuated by two short open cuts. Southbound Pennsy locomotive engineers starting their trains out of the station had to slow for a sharp curve into the tunnel, then immediately pulled their throttles open wide for a mile-long 1.34 percent grade inside. Despite the efforts of large fans at the intermediate Pennsylvania Avenue station, the tunnel was seldom free of acrid smoke — and woe to the train which stalled on the grade within it. Not only were they the worst operating problem anywhere on the Pennsylvania's eastern lines, but

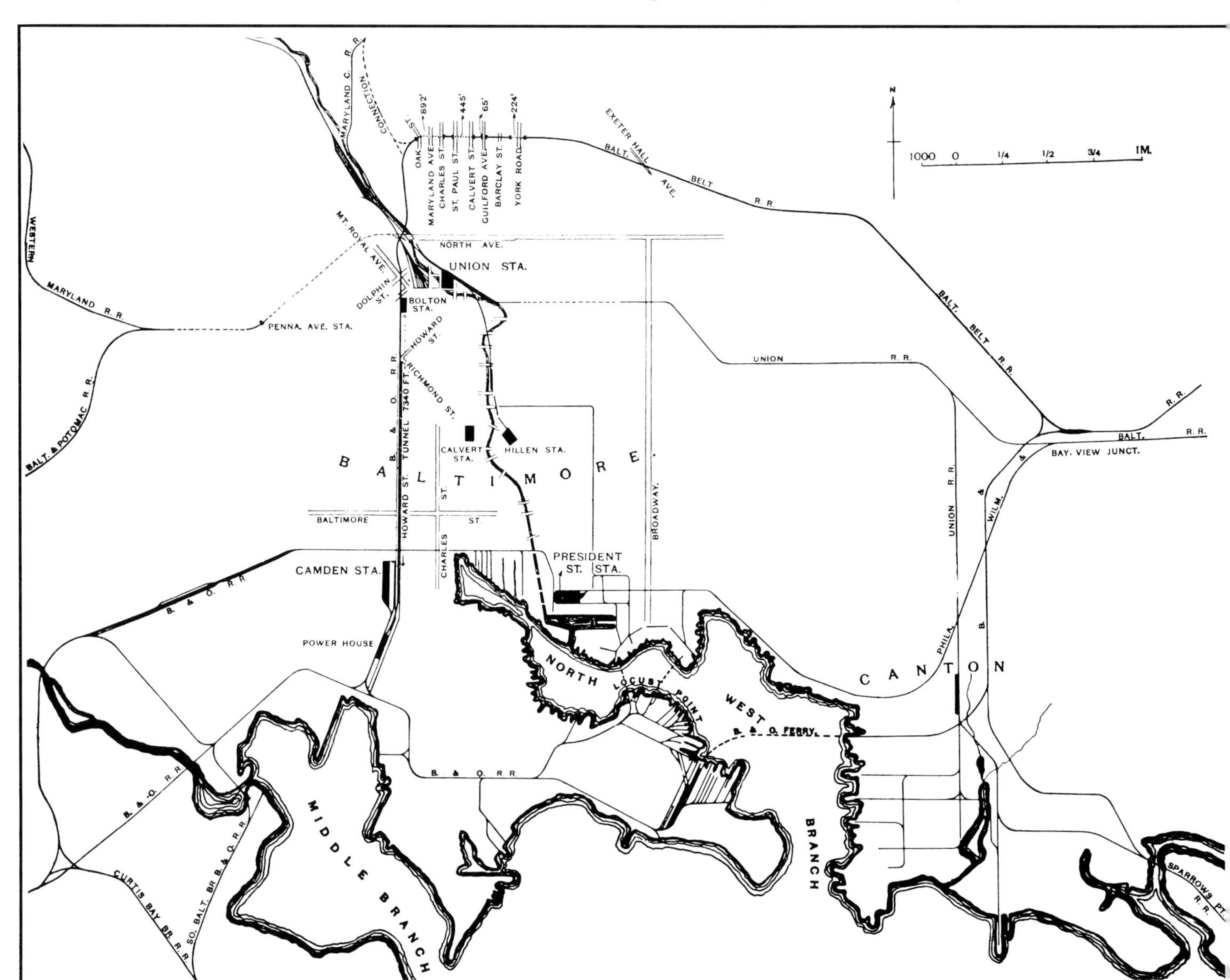

The Belt Line connected Camden Station in downtown Baltimore (at left center of map) with the original Philadelphia line at Bay View (upper right). The Philadelphia line originally began at the Canton waterfront, at the lower right.

the tunnels were the bane of the residential neighborhoods above them.

In an attempt to minimize such problems on Howard Street, the city prohibited any ventilation openings in or along the street itself; should it need tunnel ventilation, B & O was required to build elaborate chimneys on its own property, high enough to lift any smoke above the area. The Pennsy, to help alleviate its ever-worsening B & P tunnel problems, soon had to do exactly this. In 1892 it built a 160-foot-high ventilating tower at Eutaw Place; to fit into the fashionable neighborhood it was designed like an Italian Renaissance bell tower. Preferably, however, B & O had to find a way of avoiding steam power altogether in its tunnel. But at the time it began digging in 1891 it had no idea of how; nobody did, since there was no proven alternative — at least not for heavy-duty railroad service.

Fortunately, for once, the B & O's timing was right. Electric traction was in its infancy but developing rapidly. In 1885 Leo Daft had electrified a local horse car line, the Baltimore & Hampden, using dinky four-wheel 8 hp locomotives — the first successful such installation on a street railway. More significantly, just three years before B & O started its tunnel work, Frank Sprague had successfully electrified an entire city streetcar system in Richmond, Virginia. And in 1890 England's City & South London Railway opened the first electric underground rapid transit line. Its three-car trains were pulled by ten-ton single-truck locomotives with two motors totaling 50 hp. By the early 1890s, many urban horse car lines were being converted to electricity and new electric lines were being built at an accelerating pace.

But at this time the technology was still crude and confined to the light loads of street railways. And even so, not everyone was yet convinced of its practicality. Ever conservative, two Baltimore streetcar companies made the expensive mistake of building cable-powered installations between 1891 and 1893. In its early deliberations, the B & O also considered using cable power in the Howard Street tunnel, but had to conclude that it was unworkable for the loads and speeds the railroad would operate.

Electrification could have been the B & O's solution, but it was speculative. In the early 1890s electric locomotives for heavy railroad service were only theoretically possible since none existed. The only ones remotely comparable were the City & South London's ten-ton pygmies and these were remote indeed. B & O's train loads would demand more than nine times their weight and twenty times their power.

So in a sense B & O found itself back in its pioneering days with no proven designs and methods to follow. It was with immense bravery that the railroad signed an electrification contract late in 1892 with the newly-formed General Electric Company. At that point GE had not even built an

Nobody in Baltimore wanted a repeat performance of the Pennsylvania Railroad's tunnel project, completed in 1873. Its tunnels were still an operating headache 100 years later when this 1973 Penn Central scene was recorded. The Washington-bound GG-1 electric is climbing the 1.34 percent grade at the former Pennsylvania Avenue station, midway through the B & P tunnels. In steam days, fans housed above the portal attempted to clear the smoke. H. H. Harwood, Jr. photograph.

electric locomotive for heavy railroad service. It was not until 1893 that it produced a small four-wheel thirty-ton switcher, following it up a year later with a slightly larger 35-ton machine. Both worked well, but neither was heavy or powerful enough for the tunnel operation. Nevertheless, GE committed itself to building the entire system: the power generating plant, the current distribution and collection system, and locomotives capable of moving full-length passenger and freight trains up the tunnel grade at specified minimum speeds. The project was, wrote engineering historian Carl Condit, "a test case of a revolutionary form of motive power with worldwide implications."

From the outset the planned electric operation was to be different from anything before it or after. Most distinctive, the electric locomotives would haul trains only in one direction. Since the project's purpose was to solve the problem of the tunnel grade, and since the grade was entirely in the eastbound direction, it was assumed that westbound trains would not need the electric locomotives; they would simply drift downhill with their regular steam engines. Furthermore, there would be no engine changes. The electrics would tow full trains through the tunnel with their steam locomotives attached but not working. East of the tunnel, where the grades were stiffer but mostly in the open, the steamers could — and often did — pick up part of the load where needed. To cut running time and simplify the operation, the railroad originally planned to have the electrics push trains through the tunnel. The idea was discarded as being too dangerous for passenger trains; according to one contemporary trade magazine, freights still were to be pushed, but apparently that was never done either.

Undismayed by its engineering problems and its worsening finances, B & O also made elaborate plans for three new stations along the Belt Line route: one downtown inside the tunnel, a large "uptown" station north of the tunnel, and a suburban stop in Peabody Heights, at 26th Street between Charles and St. Paul Streets. But the planning was erratic and not fully thought out; in the end, only one of the three was built.

Seemingly central to B & O's thinking was that Camden Station no longer would be a viable Baltimore passenger terminal. In the words of an 1890 memorandum by John Cowen, "the [Belt Line] will be so constructed that it unfits Camden for all time for use as a general passenger depot." The reasons for such a dogmatic conclusion are unclear, but Camden did have several problems. From an operational point, its layout would be awkward once the tunnel line was opened. Camden was a stub-end terminal and its design and interior layout were oriented to the trainshed directly behind it. The tunnel's south portal was slightly east of the building and well below its level; the Belt Line's tracks continued south several blocks before they could connect with the station tracks. This meant that either an annex station would be needed to serve the tunnel tracks, or that any passenger trains using the tunnel would have to back in or out of the existing trainshed, a time-consuming maneuver that would add to terminal congestion.

Neither of these alternatives seemed worth it. Commercially, Camden was located in a slight backwater of the city and the building itself — built between 1856 and 1865 — was now architecturally unfashionable and becoming operationally obsolete. In addition, the city seemed to be growing northward away from the harbor and Camden. Optimists were predicting that eventually Baltimore's center would gravitate to North Avenue and Charles Street, more than two miles north of Camden. Although that seemed far-fetched at the time (and is even more so at present), North Avenue had become a busy crosstown thoroughfare surrounded by rapidly-growing rowhouse and townhouse developments. And, since 1873 the Pennsylvania's main Baltimore station was located on Charles Street just four blocks south of North Avenue. Rebuilt in 1886, this busy station served the Pennsy's New York-Washington route and its Baltimore-Harrisburg line, as well as trains of the Western Maryland Railway. Admittedly the Pennsy's station location originally had been forced by engineering necessity rather than any commercial foresight, but it had helped stimulate building in the area.

With all this in mind, B & O planned to move its main Baltimore passenger terminal to some site on the Belt Line near North Avenue. In 1892 an elaborate station was designed at Mosher Street just south of North Avenue and east of Mt. Royal Avenue, but property problems prevented any work. After some delay the railroad settled on a spot several blocks south, at the point where the tracks emerged from the north portal of the Howard Street tunnel. Called the "Bolton Lot," this property was an irregular block bounded by Mt. Royal Avenue and Cathedral, Preston, Brevard, and Dolphin Streets. ("Bolton" was the name of a mansion and estate in this area which had been built about 1800. The mansion itself sat on the site of the present Fifth Regiment Armory.) Bolton Lot bordered a rapidly expanding residential area, which included the townhouses of Bolton Hill and the more pretentious newly-built neighborhoods along Eutaw Place, Mt. Royal Avenue, and Mt. Royal Terrace.

Interestingly and ironically, this was to be Bolton Lot's second time around as a railroad terminal. The property (then somewhat larger) had been the early home of the Baltimore & Susquehanna Railroad, another pioneering line running north from Baltimore to York and Harrisburg, Pennsylvania. Beginning in 1832 the Baltimore & Susquehanna had built its roundhouse and shop buildings at Bolton, along with a simple two-story frame station and office located on Cathedral Street near Preston.

Subsequently, the Baltimore & Susquehanna was reorganized as the Northern Central Railway and after 1861 became a Pennsylvania Railroad satellite. In 1873 the Northern Central largely abandoned its Bolton facilities as part of the Pennsy's massive construction project which included its new Washington line, its Baltimore tunnels, and its Union Station at Charles Street. The Northern Central's trains were routed over a new alignment into Union Station, and its roundhouse and shops were relocated to a more spacious site at Mt. Vernon Yard, north of North Avenue. Afterwards the Bolton Lot had been used mostly as a coal delivery yard. In 1886, however, this too was relocated a block north in a land swap with the city, which wanted a part of the Bolton property to build Mt.

Royal Avenue and extend other streets for residential development.

But even as its tunnel construction started, B & O began having second thoughts about using its planned uptown station as its principal Baltimore passenger terminal. In addition to its Royal Blue Line passenger and freight services, all of B & O's other passenger trains would have to use the Howard Street tunnel to reach it. The company's general manager, J. T. Odell, immediately foresaw an impossible bottleneck and in March 1891 he dispatched a blunt memorandum to general counsel John Cowen. Said Odell: "The Belt Line, as being constructed, is impracticable of operation. I am beginning to think it would hardly permit the handling of the present passenger traffic, and as far as using it for freight, I will say that there are only six hours, out of the 24, when it could be used for any volume of business."

Odell made his point and the optimistic planning was scaled back. But the railroad still felt that it had to supplement, if not entirely replace, Camden Station. The solution, so it thought, was a new main passenger station inside the tunnel downtown at the intersection of Howard, Lombard, and Liberty Streets. This location was less than four blocks north of Camden and the tunnel's south portal; the distance was thus short enough so that most of the tunnel would remain relatively uncongested. This section also could be widened fairly easily to accommodate extra approach tracks. In January 1892 B & O announced its new underground central station and started work under Lombard Street to widen the tunnel.

But that was all that was ever done. The underground terminal project soon was quietly abandoned; after some hesitation, Camden was substantially altered to handle both Royal Blue Line trains and its own growing business. The station's interior was revised, new terminal tracks and a new trainshed were built on the east side of its old layout, and in 1897 a stairway, elevator, and covered platform were added to serve the tunnel-level tracks. The net effect was to reorient Camden's passenger facilities eastward to be adjacent to the tunnel. Simultaneously the freight handling facilities were grouped at the western section of the station and, beginning in 1898, an enormous warehouse was built on that side.

Perhaps it was financial pressures which killed the Lombard Street station, but it is also likely that B & O realized that it was creating another operating nightmare for itself. Once the electrification planning began, it must have become clear that all trains in and out of the underground terminal would need to be electrically-powered for what was an absurdly short distance. This would mean additional electric locomotives and crews, engine changes, and light power movements — all to move trains less than

In an early 1930s scene, a westbound train drifts under the North Avenue "twin tunnels" and crosses the one-time Northern Central main line, by then a PRR freight branch. Inside the short tunnel behind the locomotive are girder bridges over the Pennsy's tunnel, which is directly underneath. B & O Historical Society Collection, B & O photograph.

Typical of the engineering required through North Baltimore is this cut and short tunnel at Charles Street. By the time this photograph was made in July 1967, the double track had been reduced to single and the electrification abandoned. The train is an eastbound Delaware Park race special headed by a former Pere Marquette Railway E-7. H. H. Harwood, Jr. photograph.

half a mile. Furthermore, the underground station would have been confined and congested, requiring constant switching to keep it fluid. So in spite of itself, Camden survived and, in fact, slogged into the 1980s. It ended its working life as the country's oldest active urban passenger terminal, outliving B & O's own passenger service; as of 1990 it remains intact, to be preserved as part of Baltimore's new sports stadium complex. The aborted Lombard Street station survives too, a mysterious underground cavern alongside the tunnel.

The Peabody Heights suburban station was never started, but the uptown station on the Bolton Lot did finally live through all these planning gyrations. Mt. Royal station, as it was named, never would be the major terminal B & O originally had planned, but it was an important stop for all Royal Blue Line trains. It got a late start, however; property acquisition and design problems delayed the start of construction until August 1894, and Mt. Royal opened more than a year after the Belt Line itself.

In the meantime the tunnel work dragged on. Samuel Rea was an early casualty; he was forced to resign in early 1891 after overwork broke his health. Happily, he recovered, and then some. After taking a year off to rest he rejoined the Pennsylvania where he eventually was given a far more complex and challenging job: chief engineer of the vast Pennsylvania Station project in New York, with its river tunnels and electrification. In 1913 he was named the Pennsy's president.

In January 1971 a quartet of Alco S-2 switchers worked a transfer freight through the deep cut west of Edison Highway. H. H. Harwood, Jr. photograph.

The Howard Street job was difficult and delicate; it had to be done without disrupting the heavy street traffic or buildings. At one time or another a total of five intermediate shafts were used to expedite the excavation. The unstable earth regularly behaved as expected; often the daytime work would result in the street pavement sinking as much as four feet overnight, requiring a quick rebuilding job for the day's traffic. But the only major casualty was the imposing 1874 City College building at Howard and Centre Streets. In 1893 the tunnel contractors opened an underground pool of water which undermined the structure's foundations and necessitated its demolition. City College was given an even more impressive replacement building on the same site, designed by E. Francis Baldwin and completed in 1899.

The original ponderous overhead third-rail system at Guilford Avenue in North Baltimore, about 1901. Smithsonian Institution Collection, B & O photograph.

The tunnel was merely the worst part of the project, but was far from the only design or construction problem. Between the tunnel's north portal and the east rim of the Jones Falls valley, the railroad had to be threaded over, under, and around a maze of major streets and other rail lines. In quick succession the Belt Line tracks burrowed under Mt. Royal Avenue in a short tunnel, crossed the old Northern Central main line (by then a freight branch) at grade, then simultaneously passed under North Avenue while crossing over the roof of the Pennsylvania's B & P tunnel near its east portal. Both of these crossings required special engineering. Quite properly the Pennsy did not want the B & O tracks resting unsupported over its tunnel roof, so they had to be carried on a girder bridge. North Avenue crossed the Jones Falls valley on a viaduct at precisely the same location, and coincidentally with the Belt Line work the city was rebuilding it as a massive stone viaduct. B & O planned four tracks at this point; as a result the North Avenue viaduct's design had to be altered to clear the new railroad line, which would then pass under twin tunnels at the bridge's west end. As finally built, the B & O-PRR-North Avenue crossing was a three-layer stack of structures. At the same point where the B & O was tunneling under North Avenue, it was also crossing over the roof of the Pennsy tunnel on girder bridges.

The south portal of Howard Street tunnel at Camden Station, as it looked in February 1947. Directly behind the camera is the lower-level passenger platform. At the right, a pair of electric "motors" has attached to an eastbound passenger train, while a Q-4 2-8-2 with a westbound freight pokes its nose out of the tunnel. The station's baggage room can be seen above. B & O Historical Society Collection, W. R. Hicks photograph.

Immediately north of North Avenue the Belt tracks then swung across the Jones Falls valley on a six-span 503-foot girder bridge which had to be laid out on a ten-degree curve. Below the bridge was not only the stream, but

One of the original electric "motors" pauses at Mt. Royal Station in 1896 with an eastbound Royal Blue Line train and its B-14 class ten-wheeler. Note the overhead third rail suspended from the trainshed trusses. Smithsonian Institution Collection, B & O photograph.

also the tracks of the PRR's Northern Central main line to Harrisburg and the little Maryland Central. Several changes in the planned alignment had to be made through this area to satisfy the Pennsylvania and to obtain property easily. On the north side of the Jones Falls bridge the railroad had to climb a hillside out of the valley and onto higher ground in North Baltimore. The result of all this was a heartbreaking combination of curves and grades which remains an operating curse to this day.

The balance of the route to Bay View required less complex engineering design, but heavy construction work. Through the Peabody Heights-Waverly area it had to be depressed in a long cut with short tunnels where several major north-south streets crossed overhead. Another long, deep cut was needed west of the present-day Edison Highway.

Following the tunnel and line construction came the electrification installation. In keeping with the technology of the time, the system was designed to deliver 600 volts of direct current to the locomotives. A brick powerhouse was built just south of Camden Station on the east side of the tracks between Lee and Hamburg streets. This plant, powered by five E. P. Allis compound Corliss stationary steam engines, not only generated power for the railroad, but also the electric lighting for the tunnel, stations, and other company facilities in the area. B & O switched to commercially-provided power in 1914, but the powerhouse building was adapted as a car shop and survived in that form into the 1970s. It was demolished to make way for the Interstate 395 roadway.

The electric distribution and current collection system was almost comically cumbersome and seemed guaranteed to cause trouble, which it soon did. Although both overhead trolley wire and ground-level third-rail systems were then in use and proven practical, General Electric elected to design something different. Essentially it came up with a form of overhead third rail, suspended by a crude catenary system consisting of linked iron rods hung from sturdy steel truss gantries which straddled the tracks. The contact "rail" (which delivered the current to the locomotives) consisted of an inverted slotted iron trough weighing thirty pounds per foot. A metal shoe, also heavy, extended up from the locomotive's roof, fitted into the trough, and slid along (or rather was dragged) as the locomotive moved.

As originally built, the electrification began south of Camden Station and extended to the top of the grade at Huntingdon Avenue in North Baltimore, about three miles altogether. The initial operating plan was for the electric locomotives to uncouple from eastbound passenger trains while they were loading at Mt. Royal Station. Freights, however, were to be helped as far as Huntingdon Avenue.

Three locomotives ("motors" in railroad jargon) were ordered, a rather conservative number considering the potential traffic over the Belt Line. They were big, even by steam locomotive standards, weighing 98 tons with four pairs of 62-inch drivers in a two-section articulated frame. More than half their weight was in their four enormous quill-drive electric motors, each of which turned out 360 hp. With a total of 1,440 hp, the new B & O electrics could produce 49,000 pounds of tractive effort — 27 percent greater than the railroad's most powerful 2-8-0 steam freight locomotive. Their original specifications required them to pull 500-ton passenger trains upgrade through the tunnel at 35 mph and 1,500-ton freights at 15 mph.

The last survivor of the original electric roster, No. 2, was preserved and later relettered as No. 1 for exhibits. Here it sits at Halethorpe, Maryland after the 1927 "Fair of the Iron Horse." Tragically, this priceless relic was scrapped in 1935. T. B. Annin photograph.

By spring of 1895 most of the basic Belt Line construction work was completed, although the electrification was still under way and no locomotives had been delivered. Mt. Royal Station also was unfinished and Camden's rebuilt facilities were incomplete or, in the case of its new lower-level platform, not yet started. Regardless, B & O was anxious to get into operation as quickly as it could. On May 1, it sent its first scheduled passenger train through the tunnel and across the new line. The honor went to No. 514, the overnight Washington-New York run. By 1 p.m. the following day, a total of 26 trains had used the new Belt Line. There seems to be no record of how bad the smoke was, but whatever the problems, pure steam passenger operations continued for almost two months. (Freight continued moving across the harbor to Canton for the time being.) With all passenger trains now using the Belt Line, the segment of the original Philadelphia main line on the east side of the harbor between Canton and the Belt Line junction at Bay View became a freight-only branch.

The first electric locomotive arrived from GE's Schenectady, New York plant in early June 1895, and on June 27th it made its first trial runs: the first electric to run on an American steam railroad. Coincidentally or not, it was numbered "1" on B & O's roster. Extensive testing and adjustments followed through September 1895; for most of this time the single locomotive apparently was tried mainly on freights while the bulk of the passenger runs continued to be steam-hauled. During one later test the electric "motor" hauled a 44-car 1,900-ton freight and three steam locomotives through the tunnel at 12 mph, turning out an astonishing 63,000-pound starting tractive effort. Another time it was run light through the tunnel at an exhilarating 61 mph.

The second "motor" arrived disassembled in late November and was completed and put in service; the third finally arrived in May 1896 and the operation was considered fully electrified. According to official company reports the Canton carferry operation ended in August 1895, although at least one B & O historian believes that some freight may have been carfloated across the harbor until 1903 when B & O added four more electrics specifically designed for freight use. (It should be noted that freight business on the east side of the harbor itself was growing, primarily from the new Sparrows Point steel complex and industry in Canton.) The carferry *John W. Garrett* was sold in 1899 to the Norfolk & Southern Railroad (later Norfolk Southern) and used to ferry trains across Albemarle Sound in North Carolina until 1910. It then went to the Frisco system for Mississippi River service and finally ended its life as a barge.

Passenger facilities in Baltimore were in some disarray for the first several years of the Belt Line's life. Uncertainty over Camden's status had delayed building the low-level platform facilities and, until 1897, all Royal Blue Line trains backed in or out of the upper-level trainshed. The last major piece of the Belt Line project, Mt. Royal Station, finally opened its doors September 1, 1896 after two years of construction. Its design, too, presented problems. The site was difficult since the tracks came out of the tunnel below street level and then plunged under Mt. Royal Avenue a block north. After debating the merits of a two-level station, B & O decided that it should be built entirely at track level. Thus the block had to be extensively excavated and regraded, placing the building in a wide artificial hollow below street level.

This location detracted somewhat from Mt. Royal's visual impact, but unquestionably it was an impressive structure. It had been designed by E. Francis Baldwin, the Baltimore architect who had created so many outstanding B & O buildings in the 1870s and 1880s. Mt. Royal was to

Photos on this page and on opposite page, top, *show Mt. Royal inside and out, in its earlier days. The spectacular building remains externally unchanged today, although the spacious and airy waiting room has been subdivided into galleries and offices. Exterior and interior views: B & O Historical Society Collection, B & O photographs. Porte-cochere view: F. A. Wrabel Collection.*

The north end of Mt. Royal's trainshed, looking toward the Howard Street tunnel. Taken as the station was being completed, the photograph shows the catenary suspension of the original overhead third rail. Smithsonian Institution Collection, B & O photograph.

Fifty-six years later, in 1952, the same scene was almost unchanged as the New York-bound Capitol Limited roars out of the tunnel. A four-track section started here, its switches controlled by "RM" tower at the far right. James P. Gallagher photograph.

be Baldwin's last, largest, and finest B & O station — and unlike anything he had done earlier for the company. Architectural fashions were changing, and Mt. Royal emerged in a grandiose but hybrid Romanesque style. Originally planned as an asymmetrical Picturesque Romanesque Revival building, it was to have a tower at one corner and a rough stone finish. But midway through the design process it was changed to the smoother, more Classical Second Renaissance style then coming into vogue.

As finished, it was a symmetrically-designed two and one half story building with a wide 250-foot frontage. A pair of three-story end wings flanked the center section, and a soaring 150-foot-high central clock tower helped lift the station from its sunken site and make it visible to the surrounding neighborhood. It had massive walls of Port Deposit granite trimmed with Indiana limestone, upper story windows resembling those of a Florentine Renaissance palace, and a red tile roof. Inside the station's central section was a 125-foot-long vaulted waiting room two stories high; typical of the time, it was informally divided into "general" and "ladies" sections, with a central ticket office and a small restaurant.

In front, a curving driveway led downhill into the "hollow" from Cathedral Street, past lawns and gardens to the large stone *porte-cochere* in front of the tower. Covered stairways and walkways decorated with ornamental ironwork also connected the station with Mt. Royal Avenue and with Brevard Street. To reach the trains from the airy waiting room, passengers walked through rear doors which opened into a skylighted trainshed, separated from the station by delicate and ornate iron gates. The shed, 400 feet long, abutted the Howard Street tunnel portal and spanned four tracks which immediately narrowed to two as they plunged into the tunnel. In 1936 the trainshed's north end was extended to Mt. Royal Avenue to accommodate longer trains.

In short, Mt. Royal was everything that Camden was not — spacious, supremely fashionable, aesthetic, and well-matched to its genteel surroundings. Its planners were foresighted to the extent that Baltimore's "better" residential areas continued moving north and northwest, and Mt. Royal was well situated to serve these areas. Throughout its life the station remained a more convenient location than Camden for many suburban Baltimoreans, and all Philadelphia and New York trains stopped there. But the city's commercial center never strayed far from the old axis at Baltimore and Charles Streets, even after the 1904 fire wiped out most of downtown, and Mt. Royal served out its life essentially as a suburban station. In its way, the monumental station was as symbolic of the Royal Blue Line as the ornate rolling stock and graceful high-drivered locomotives: beautiful, luxurious, and underutilized.

Mt. Royal's opening was the last move in making the Royal Blue Line the full competitive equal of the Pennsylvania for the New York-Washington passenger trade. In fact, the B & O route was now superior in some ways. Its trains could get through Baltimore as quickly as the Pennsy's, but much more cleanly. Furthermore, B & O offered two Baltimore station locations, central and suburban, while the PRR's single station near North Avenue was, and always would be, less convenient to the city's downtown. Elsewhere along the route, running times already had been made generally equal, although the Pennsy still was slowed some by its backup movement at Broad Street Station in Philadelphia.

The new Belt Line opened other significant commercial opportunities too. Although built primarily for the

Philadelphia and New York business it also gave B & O direct rail access to the eastern side of Baltimore's harbor. Through much of the first half of the 20th Century, this area steadily industrialized and became one of B & O's largest freight traffic sources in Baltimore. By the 1960s, in fact, the Belt Line was far busier with local industrial traffic than it was with Philadelphia/New York freight movements.

The cost of all this was something else. Adding the Belt Line to the Philadelphia Division construction, the Staten Island facilities, and all the new locomotives, cars and marine equipment pushed the price of B & O's New York entry to at least $35 million. The Belt Line project alone ultimately cost about $7 million — mile for mile, the most expensive ever undertaken by the company. On February 29, 1896, as the final touches were being put on the Belt Line, B & O sank into receivership. Garrett's New York dream had been a major cause, although certainly not the only one. But the ironies were many. As B & O's financial crisis accelerated in 1895 and the railroad cut back wherever it could, Mt. Royal Station became its last major active construction project and probably its most pretentious and least necessary one. When completed it cost $300,000.

More relevant, while B & O had been devoting the bulk of its borrowed resources to the prestigious New York route, its main line coal and merchandise trains were struggling around tight curves and over heavy grades laid out in the 1830s and 1840s. Its obsolete physical plant ballooned expenses and slowed service, all of which also helped to bring on the receivership. By the turn of the century, B & O was forced to catch up quickly with its competitors by rebuilding the entire railroad system, but too much was needed all at once. Remembering the Belt Line's $7 million cost, it might be noted that a later B & O president, Daniel Willard, had to wait until 1911 before he could spend $6 million to build the Magnolia Cutoff on the main line in the Potomac Valley, a desperately needed project to clear the system's worst bottleneck at its busiest spot.

Whatever its logic, the B & O now had a truly first-class railroad to Philadelphia, and with the Belt Line electrification it had reclaimed its pioneering reputation. Now it intensified its efforts to run a railroad worthy of the money spent on it.

Turn-of-the-Century Splendor and Speed

1898-1918

It was the best of times — or so it would seem in the years afterwards. In the late 1890s and early 1900s the Royal Blue Line reached a peak of elegance and physical perfection. And although scheduled speeds would continue to improve through the 1930s, the *image* of speed would never be stronger or more dramatic, with handsome, high-wheeled ten-wheelers and Atlantics rolling their short trains at 90 mph and more. Services and schedules were as close to the Pennsylvania's as they ever would be. Briefly the world looked bright, until the Pennsy delivered its ultimate blow.

It was a frantically busy time too, with substantial plant rebuilding, new facilities, a rapid succession of new power, new services, and corporate turbulence — all happening more or less at once.

Concurrent with the completion of the Baltimore Belt project, B & O began supplementing its almost-new M-1 class 4-4-0s with larger power. Pushed by the Pennsy, which had begun installing its fast, heavy D-16 class Americans on the New York route in 1895, B & O found that many of its ten-year-old bridges could not take the axle loadings

Resting or rolling, B & O's new ten-wheelers were a handsome breed. B-14 No. 1322 waits with a Washington train at Camden Station's lower level platform. Smithsonian Institution Collection, B & O photograph.

A later B-17 in flight with a three-car train of Royal Blue Line equipment somewhere outside Philadelphia about 1902. H. H. Harwood, Jr. Collection.

of a heavier 4-4-0 of its own. So between 1896 and 1897, Baldwin built ten new high-speed ten-wheelers, a wheel arrangement which allowed greater weight by spreading the load over more axles. Like their M-1 predecessors, B & O's new B-14 class 4-6-0s rolled on 78-inch drivers, but otherwise they were 35 percent heavier and could turn out almost forty percent more tractive effort.

Quickly afterward came the pinnacle of luxury: the *Royal Limited*. Built to challenge the competition's *Congressional Limited* (which actually had been operating since 1882), the *Royal Limited* was a five-hour, all-Pullman parlor car train aimed at the first-class political and business trades. It was more than just a fancy train. With its overpoweringly opulent decor and its unashamed imagery of royalty and wealth, it was a rolling symbol of the late Victorian world. By the time the *Royal Limited* made its first run on May 15, 1898, however, it was a world which was heading into deep twilight.

Initially a set of twin four-car trains, each *Royal Limited* consisted of two parlor cars, a diner, and a cafe'/smoker/observation car. The Pullman Company provided four wide-vestibule parlor cars which were specially fitted for the train. Carrying names like *Queen, Empress, Czarina,* and *Countess,* they surely lived up to their names. Interior decor was Victorian opulence run rampant. Only the B & O's own description does it justice: "The parlor cars are superbly finished in vermilion wood, with an inlay of Persian design. The ceiling is of royal blue, and the upholstery of the same color, except in the ladies' toilet, which is of dark olive green, the ceiling decorated to correspond. The drawing room is most beautifully finished in green and gold. The general design of the main parlor is Persian, whilst in the drawing room and ladies' toilet the design is renaissance. A beautiful effect is given both to the exterior and interior of the cars by oval windows, with opalescent glass placed in the toilet rooms and passageways." Window curtains and chair cushions were fringed and tasseled, and the spacious "ladies' boudoir" had a large, ornate mirror and a bookcase.

The equally sumptuous diners (actually diner-baggage combines) named *Waldorf* and *Astoria* were Pullman-built but B & O-operated. Their full-course meals were served on "the daintiest china and cut glass," the china decorated in a specially-chosen shade of blue made only in Austria. (A year later these full diners were supplemented by a pair of unusual cafe'-diners — *Savoy* and *Manhattan* — with center kitchens, giving travelers a choice of either elaborate or quick meals.) One of the parlor-observation cars, *Jupiter, Mercury,* or *Neptune,* brought up the rear; these included wicker lounge chairs, a writing desk, and a buffet

Shortly after the opening of Washington Union Station in 1907, a four-car Royal Limited waits to carry some distinguished but now-forgotten group to New York. Smithsonian Institution Collection, B & O photograph.

section. Unencumbered by modesty, B & O advertised its *Royal Limited* as "The Finest Day Train in the World." It was probably correct.

Royal Limiteds left both New York and Washington in mid-afternoon and, exulted a typical 1899 ad, covered "Five States in Five Hours." Afterwards, schedules and equipment varied, but the *Royal Limited* name and deluxe service survived until 1916. Apparently B & O had difficulty sustaining it as an exclusively all-parlor train, however; for one year between 1900 and 1901, and permanently after 1911, coaches were added to its consist. (Significantly, the Pennsy's *Congressional Limited* remained all-parlor until the Depression.) Beginning in December 1916 the *Royal Limited* was renamed the *National Limited* and began carrying through cars for B & O western points. By then the charm of Old World royalty had vaporized in the blood and mud of the First World War.

Shortly before the *Royal Limited's* inauguration in 1898, B & O also attempted to improve its Manhattan passenger terminal. Traditionally, Royal Blue Line passengers rode the Jersey Central's ferries to and from Liberty Street in lower Manhattan, almost directly across the Hudson from the Jersey City terminal. Liberty Street (now next to the World Trade Center complex) was reasonably close to New York's financial district, but the CNJ terminal lacked direct access to public transportation. The nearest rapid transit line, the 9th Avenue elevated, was a block away; the other north-south "els" and the major street railway lines were even farther. (The Pennsylvania's situation was similar, although it offered more ferry routes and terminals. At the time, Pennsy ferries operated from Jersey City to Cortlandt Street, adjacent to Liberty Street, and to Desbrosses Street, 23rd Street, and Brooklyn.)

With its acquisition of the Staten Island Rapid Transit, B & O had inherited a better located New York ferry terminal. SIRT's ferries operated to Whitehall Street at the lower tip of Manhattan; its terminal, locally called South Ferry, was the hub of New York's rapid transit system. All four north-south elevated lines terminated here with covered walkways to the ferry terminal, and the Broadway cable railway route and other major surface lines ended on the street outside.

As early as 1893 B & O began planning to route some Jersey Central ferry services to Whitehall Street. To do so,

Photos above and right: *Pullman parlor car Queen and parlor-observation Mercury were typical of the Royal Limited's early consists. Both were turned out by the Pullman Car Works in 1898. Smithsonian Institution Collection, B & O photographs.*

CNJ ordered two new boats specially designed for the Whitehall Street slips — the *Easton* and *Mauch Chunk*; these also were the railroad's first double-deck and propeller-powered ferries. After considerable delay the Whitehall Street service finally started July 19, 1897, giving "Blue Line" passengers a choice of two Manhattan landing points. Sadly, it soon led to disaster. On June 14, 1901 the *Mauch Chunk* collided with the Staten Island's *Northfield* at the South Ferry slip. The *Northfield*, carrying 995 people, sank immediately. Miraculously only five passengers were drowned, but the accident helped build pressure for public ownership of the Staten Island ferry service. In October 1905 the city took title to the Whitehall Street terminal and the Staten Island boats, and all railroad-owned services ceased.

But by then the Jersey Central and its partners had changed strategy. On June 25, 1905 CNJ dropped the South Ferry service and simultaneously began operating to a new midtown terminal at the foot of West 23rd Street. At that time 23rd Street was briefly the commercial hub of the rapidly growing city, near many of the most fashionable retail stores and hotels. And although the 9th Avenue elevated station was two blocks away, several streetcar lines ended at the ferry terminal, including a direct service to Grand Central Station on 42nd Street. Indeed, the newly-built 23rd Street ferry terminal complex was a bustling magnet for most railroads which terminated on the New Jersey side of the Hudson. In addition to the Jersey Central (which, of course, also carried B & O and Reading passengers as well as those of the Lehigh Valley), ferries of the Lackawanna, the Erie, and the Pennsylvania ran to 23rd Street. The new ferry house promptly burned down that December but was soon rebuilt. The city's center of gravity kept moving north, soon leaving 23rd Street in a backwater, but the ferry terminal remained as an alternate route for CNJ, B & O, and Reading riders until November 14, 1941. The original Liberty Street service doggedly outlasted ev-

An eastbound Royal Blue Line train powered by a B-17 ten-wheeler hurdles the Brandywine in Wilmington about 1901. F. A. Wrabel Collection.

And at about the same time, a westbound edges off the Susquehanna bridge at Havre de Grace, Maryland. Smithsonian Institution Collection, B & O photograph.

erything, including the end of B & O's New York passenger operations; it finally succumbed in 1967.

Back on the rails, motive power reached a peak of fleetness and grace which was never surpassed. Pushing for more power and speed, both the Pennsylvania and the three Royal Blue Line partners once again reequipped their passenger locomotive rosters with new designs.

The Reading-Jersey Central combination was first to move; in 1899 both railroads introduced 4-4-2 Atlantic camelbacks on their joint Philadelphia-New York route. These regal racehorses rolled on 84-1/2-inch drivers, the largest yet used on the line, and originally were built with Vauclain compound steam distribution systems. Over the next several years the two companies added successively more refined versions of the type, culminating with the Reading's home-built P-5 class Atlantics fitted with unheard-of (and never repeated) 86-inch drivers. An odd contrast of grace and ugliness, the high-wheeled hump-backed center-cab CNJ and Reading Atlantics were among the fastest and most memorable steam power ever to roll Royal Blue trains. Through a near-miracle (its scrapping had already begun), one sample was saved: the Jersey Central's No. 592, a 1901 Brooks Locomotive Works product. With its original 85-inch drivers long ago reduced to a more sedate 79 inches, and now crowded inside the B & O Railroad Museum at Baltimore, No. 592 still looks eager to get out and run fast.

Hardly one to be left behind mechanically (or any other way), the Pennsylvania introduced Atlantics on its

Photos left and on opposite page, lower right: *By the early 1900s Atlantics like these were wheeling B & O trains between Philadelphia and Jersey City, sometimes at scorching speeds.* ***This page:*** *Reading's 326, a 1900 Baldwin product, proudly poses at Philadelphia in 1904. Originally a Vauclain compound, it had just been rebuilt. D. A. Somerville Collection, F. H. Somerville photograph.*

New York-Washington trains in 1901: slim, smooth-boilered end-cab E-2 and E-3 classes which, with 80-inch drivers, could move every bit as fast. It was an E-2 which, by legend at least, set a record of 127.1 mph in Ohio in 1905.

Still hampered by bridge axle loading restrictions, B & O had to use only ten-wheelers between Washington and Philadelphia. In 1901 it added nine more, giving it a total of nineteen fast 4-6-0s exclusively assigned to Royal Blue Line through trains. The newcomers, also Baldwin-built, were Vauclain compounds and classed as B-17s, but otherwise were similar to the earlier B-14s. (The Vauclain compound system was an attempt to increase efficiency by using the engine's steam twice.) The two ten-wheeler groups remained the backbone of B & O's "Blue Line" services for almost a decade, turning in highly respectable running times over the curving, bridge-punctuated Philadelphia Division. The Vauclain compound system, however, was far less successful, and in 1905 the B-17s were rebuilt with conventional single-expansion cylinders and slide valves. Reading and CNJ also soon "simpled" their Vauclain compound Atlantics, as did all other railroads which tried the system.

By the early 1900s Royal Blue Line service had settled down to eight New York-Washington trains each way — a pattern which, with occasional additions or subtractions, would remain essentially the same for almost fifty more years. Despite the Baltimore Belt Line and the efforts of the new ten-wheelers and Atlantics over the route, five hours surprisingly remained the fastest time. In 1904, for example, two Royal Blue trains each way were scheduled for five hours; most others took between five hours fifteen minutes and five hours fifty minutes. (Under no real pressure, the overnight trains crept along in seven hours.) Competitive Pennsylvania times were comparable. Although the Pennsy scheduled more New York-Washington trains — fourteen each way, not counting through services to the South and to Boston — only its all-parlor *Congressional Limited* made the trip in five hours.

From an 1898 advertisement. H. H. Harwood, Jr. Collection.

But although the B & O and PRR schedules looked much the same, their services and markets had some basic differences which remained over the life of the competition. Thanks to its earlier start and generally better location, the Pennsy already was evolving into the mass carrier over the

Familiar to B & O Museum visitors, the Jersey Central's No. 592 is shown as new at Jersey City with its 85-inch drivers. Smithsonian Institution Collection, C. B. Chaney photograph.

B & O train 507, the New York-Chicago Special, accelerates out of Jersey City in 1910 behind Reading No. 346, a 1906 P-5 class Atlantic. Smithsonian Institution Collection, C. B. Chaney photograph.

route. Most of its trains were oriented strictly to business between New York and Washington and their intermediate points: notably Newark, Trenton, Philadelphia, Wilmington, and Baltimore. In addition, its line formed the central segment of a much longer travel corridor stretching from Boston into the South. With its control of the Washington gateway and a trainferry operation across New York Harbor, the PRR also operated a wide range of inter-regional through services. Solid passenger trains and blocks of cars ran between New York and points on the Southern Railway, the Chesapeake & Ohio, the Atlantic Coast Line, and after 1900 the Seaboard Air Line Railway. To the northeast the Pennsy ran through Washington-Boston trains via its carferry connection with the New Haven. All these services also were funneled over the increasingly busy New York-Washington rails.

B & O, on the other hand, was precluded from operating through southern or New England passenger services, although it solicited interterritorial transfer business between stations at New York and Washington. B & O also worked with the New England-New York steamship lines and the Chesapeake Bay boats between Baltimore and Washington and the South. But it did use its New York route as the eastern end of its main line passenger services to western points such as Chicago, Cincinnati, St. Louis, and Pittsburgh. Except for a few trains like the *Royal Limited*, typical Royal Blue Line trains carried through western cars which were switched in and out of main line trains at Washington or Baltimore. In doing so, B & O tried to compete directly with the Pennsylvania's and the New York Central's east-west limiteds at New York. Fairly consistently over the years, at least two Royal Blue Line trains each way offered a New York-Pittsburgh-Chicago service, and two carried through cars between New York and Cincinnati-St. Louis. For most western destinations, B & O's route through Washington and the switching between trains made it longer and slower than the Pennsy and Central. Adept at making a liability into a virtue, B & O advertising proclaimed "All trains via Washington...with Stop-Over Privilege" and featured a drawing of the Capitol dome. The dome symbol became so solidly linked with the railroad that in 1937, abstracted from its original message, it was made B & O's official corporate logo.

But B & O's turn-of-the-century president, John Cowen, had bigger worries than passenger competition. Having nursed the company out of its brief receivership by 1899, he now faced the unthinkable — a control takeover by the B & O's mortal enemy, the Pennsylvania. In addition, although a growing business was beginning to help its financial problems, the railroad required massive rebuilding to cope with the traffic and heavier equipment. The Garrett legacy, including the diversion of so much capital to the New York adventure, had left the B & O physically far behind its competitors.

The competition could move fast too. On the Pennsy's New York-Washington line an E-3/E-2 doubleheader rolls an eight-car train in 1912. Smithsonian Institution Collection, C. B. Chaney photograph.

The Royal Blue Line

CENTRAL RAILROAD OF NEW JERSEY
PHILADELPHIA & READING RAILWAY
BALTIMORE & OHIO RAILROAD

FINEST SERIES OF PASSENGER TRAINS IN THE WORLD

SPLENDID ROYAL BLUE LINE COACHES. PULLMAN DRAWING-ROOM PARLOR AND SLEEPING CARS. UNEXCELLED BALTIMORE & OHIO DINING CAR SERVICE

SCHEDULES EASY TO REMEMBER

FROM WASHINGTON

BALTIMORE & OHIO R.R. TO BALTIMORE
PULLMAN SERVICE

"EVERY HOUR ON THE HOUR" FROM WASHINGTON TO BALTIMORE
DAILY, EXCEPT SUNDAY. RETURNING IN LIKE MANNER
"EVERY OTHER HOUR on the ODD HOUR" to PHILADELPHIA and NEW YORK
ADDITIONAL TRAINS AT 8.00 P. M., 11.30 P. M. AND 2.57 A. M.

FROM NEW YORK

"EVERY HOUR ON THE HOUR" FROM NEW YORK TO PHILADELPHIA
DAILY, EXCEPT SUNDAY. RETURNING IN LIKE MANNER
ADDITIONAL TRAINS FROM NEW YORK AT 7.30 P. M., 12.15 NIGHT AND 4.30 A. M.
"EVERY OTHER HOUR on the EVEN HOUR" to BALTIMORE and WASHINGTON
ADDITIONAL TRAINS AT 7.00 P. M. AND 12.15 NIGHT

(1904) F. A. Wrabel Collection.

And unexpectedly, the Philadelphia line itself had to be rebuilt. Although only fourteen years old in 1900, it already was showing some weaknesses. Most troublesome, its many pin-connected Pratt truss bridges and iron trestles could not handle the latest locomotive designs without compromising speed. Less costly but equally pressing was the need to replace the scattered patchwork of freight yards and the locomotive facility at Philadelphia.

Before Cowen had much chance to do anything, the Pennsylvania, now led by the aggressive Alexander J. Cassatt, had bought control of the B & O. (This was the same Cassatt, incidentally, who had helped engineer the PW & B coup in 1881.) In 1901 Cassatt demoted Cowen back to general counsel and installed a PRR officer as B & O president, the dynamic Leonor F. Loree. But to everyone's surprise it was a happy move for the battered B & O. An individualist and a builder, Loree wasted no time. He laid out a system-wide improvement program and instantly began implementing it, picking up some projects already haltingly begun and adding many more. As much as B & O's resources would allow, Loree aimed at duplicating the Pennsylvania's own heavy rebuilding program being pushed by Cassatt.

Loree's reign was destined to last less than four years, and soon afterward antitrust pressures made it politically prudent for the Pennsy to back away from its B & O control. But his program and projects were picked up and completed by his successors, Oscar Murray and Daniel Willard. By the end of the first decade of the 20th Century, the Philadelphia Division's bridges had been almost entirely replaced and it had its new Philadelphia yard and engine terminal. This time the job lasted; the same structures and facilities are still serving the railroad into the 1990s.

The bridge program was started in 1900 but went slowly for the first few years because of so many projects elsewhere on the system — and also because normal full train schedules had to be maintained through the work. An extreme example of this problem was at Swan Creek, Maryland, between Aberdeen and Havre de Grace where a set

Baltimore & Ohio Dining Car Service

ROCKAWAYS ON HALF SHELL
POTAGE A LA REINE GREEN TURTLE CLEAR
CELERY SALTED PECANS OLIVES
BAKED COD STUFFED WITH OYSTERS
CUCUMBERS
LOBSTER A LA AMERICAINE WHIPPED CREAM PUFFS
ANGELS ON HORSEBACK
BAKED SMITHFIELD HAM, WINE SAUCE
ROAST BEEF, AU JUS
MASHED POTATOES STEWED TOMATOES
ROAST YOUNG TURKEY, CRANBERRY JELLY
CANDIED SWEET POTATOES ASPARAGUS ON TOAST
VENISON STEAKS A LA CHASSEUR
ROAST PRAIRIE CHICKEN, CHESTNUT DRESSING
(Water Cress)
Champagne Punch
LETTUCE AND TOMATO SALAD, FRENCH DRESSING
ENGLISH PLUM PUDDING, HARD OR BRANDY SAUCE
PLOMBIER ICE CREAM ASSORTED CAKES
STRAWBERRIES AND CREAM FRUIT JELLY
ICED MALAGA GRAPES
EDAM AND ROQUEFORT CHEESE
BENT'S CRACKERS
COGNAC

The Drinking Water is from the Spring at Deer Park, Md.

MEALS $1.00

CAR 1008

(1904) F. A. Wrabel Collection.

B & O B-17 No. 1331 with eight wooden cars shortly after its rebuilding in 1905. Smithsonian Institution Collection, B & O photograph.

of track water pans was located. Trains could not pick up water if they were slowed while the bridge was being replaced. Thus when the work was finished in 1906, the job of removing the old truss span and dropping in a new girder bridge was done between trains in four minutes.

The long Susquehanna bridge, of course, was its own special problem. Work started in 1907 on a new double-track steel bridge to replace the old single-track structure on the same alignment, using the original piers. This time, though, the attempt to run trains while rebuilding ended in disaster. On September 23, 1908 one of the eastern spans collapsed under a coal train, dumping several loaded hopper cars into the river. Afterwards a detour route was worked out over the closely-parallel Pennsy line, using its recently replaced Susquehanna bridge. Connecting tracks were built at Swan Creek on the west end and Perryville on the east. While no accounts seem to survive, the detour must have been slow and difficult. The Swan Creek connection was laid out in the opposite direction from the normal movement, requiring a backup move on or off the Pennsy. At Perryville, B & O trains had to use the Pennsy's Columbia & Port Deposit freight branch a short distance, then negotiate a steep grade to the B & O main line at Aiken which at that point was about fifty feet above the C & PD line. (Helper engines were usually needed for eastbounds over this short stretch.) In between the two points, B & O trains undoubtedly had to wait to get onto the busy Pennsy main line.

The Susquehanna River ordeal finally ended January 6, 1910 when B & O's fully-rebuilt bridge was put in service. The configuration of the new bridge was essentially like the old, although it employed additional piers with shorter and more numerous deck truss spans (eleven versus the original seven); also, a series of 23 girder spans substituted for the former trestlework over Garrett's Island. The detour track connection between Swan Creek and the Pennsy's Oakington station was removed in 1918, but the tortuous Perryville-Aiken link remains in use in 1990.

On August 3, 1908 trains were still running over the Susquehanna bridge while rebuilding was in process. Seven weeks later a collapse forced a fifteen-month rerouting over the Pennsy's paralleling bridge. B & O Historical Society Collection, B & O photograph.

Finished in 1909 and opened in early January 1910, the new Susquehanna bridge was still in service in 1990. In this 1968 view, a westbound freight approaches the Havre de Grace end. H. H. Harwood, Jr. photograph.

Most other bridges were replaced in kind with similar but heavier structures, although an earth fill took the place of the long, high Big Northeast Creek trestle between Leslie and Childs, Maryland. The last large piece of the rebuilding project, the Brandywine bridge at Wilmington, also opened in 1910. In this case the old six-span deck truss bridge was replaced by a graceful seven-arch stone Pennsylvania Railroad-style viaduct on a slightly different alignment. Bypassed but not demolished, the original 1885 bridge remained in place and in 1920 was rebuilt as the Augustine Cutoff highway bridge. It survived in that form until 1980 when its superstructure was replaced. Today the twice-rebuilt Augustine bridge still soars over the Brandywine supported by the original stone railroad piers.

Amazingly, two large original Royal Blue Line truss bridges survived in active service in 1990. With the Baltimore Belt Line's opening in 1895, the old main line between the Canton ferry pier and the Belt Line junction at Bay View was demoted to a freight branch. On this section were (and still are) two unusual skewed pin-connected double-intersection Pratt through trusses, one crossing the PRR (now Amtrak) main line at Bay View, the other crossing the original PW & B main line (currently a Conrail freight branch) a short distance to the south near Gough Street in Highlandtown. Long since single-tracked, these two ancient spans currently carry a dense traffic of coal, steel products, auto parts, new autos and trucks, and trailer / container shipments.

Locating a site for a central Philadelphia freight yard was difficult; in fact, no solution was completely satisfactory. Ideally, any major yard should be laid out parallel to the main line to allow direct movements in and out. In Philadelphia this turned out to be impossible; in effect the railroad was painted into a corner. B & O's main line hugged the Schuylkill River on one side and was hemmed in by the city on the other. Nor was there any satisfactory yard site on the west side of the river, where it crossed over the Pennsylvania's Washington main line and passed through an urbanized area.

As the least of all evils, it was decided to expand the small facility at East Side, on the east bank of the Schuylkill where the Delaware River branch left the main line. The yard itself and the new engine terminal were built along the branch's right of way immediately south of the main line junction, where there was flat open ground. Construction commenced in 1906 and was completed by 1908. Operationally it was a difficult, awkward layout. The yard and branch could be entered from only one direction, since there was no room for a wye junction and the direction of

A lesser example of the early 20th Century rebuilding program was the Gunpowder River bridge near B & O's Bradshaw station in Maryland. Here it supports an eastbound train in 1977. H. H. Harwood, Jr. photograph.

entry was from the east rather than the west. This meant that freights from Baltimore to Philadelphia had to pull past the junction and yard leads, then stop and back into the yard. Trains to Baltimore struggled out the opposite way, backing out of the yard onto the main line before heading west. The curse continues today, although over the years B & O gradually reduced it by transferring as much freight switching as possible to Wilsmere Yard near Wilmington.

At Wilmington, this sturdy stone viaduct replaced the original Brandywine bridge, which can be seen behind its successor in this 1975 photograph. H. H. Harwood, Jr. photograph.

Still serving into the 1990s was this original 1885 truss bridge over the PRR at Bay View in Baltimore. Double-tracked when built, it was single-tracked after the old Canton line became a freight-only branch. The photograph dates to 1978. H. H. Harwood, Jr. photograph.

But for better or worse, East Side has contiually served as B & O's primary Philadelphia yard.

Adjacent to the freight yard and finished at the same time was the new engine terminal, which handled all B & O power for the Philadelphia area. In 1908 the now-rickety "temporary" 58th Street engine terminal (by now 22 years old) was closed and its machinery transferred to the modern 25-stall East Side roundhouse. East Side survived the demise of steam and served well into the diesel era before it burned down in October 1969.

North of Philadelphia the Reading made its own contribution to help speed up Royal Blue Line operations. In 1906 it opened its New York Short Line, a 9.6-mile bypass route from Cheltenham (in northern Philadelphia) to Neshaminy Falls, Pennsylvania where it rejoined the original Bound Brook line. The New York Short Line allowed B & O passenger trains and any through freights to avoid the more roundabout, steeply-graded, and congested suburban route through Jenkintown. At the New York end of the route the Jersey Central rebuilt its long Newark Bay trestle between 1905 and 1906 by installing a rolling lift bridge to clear the growing water traffic faster.

In Baltimore, the Belt Line electrification was put into what would be its final form. In 1901 the electric operation, still using the cumbersome and troublesome overhead third rail, was extended a mile eastward from Huntingdon Avenue to the Waverly section of north Baltimore. By then this section of town was becoming a tightly built-up residential area and smoke was a nuisance. A year later General Electric admitted failure with the overhead third rail which, among other things, suffered from severe corrosion inside the tunnel. The structure was dismantled and replaced with a more conventional ground-level third-rail system which lasted until the electrification itself was abandoned fifty years later. A year after that, in 1903, four new electric "motors" arrived to supplement the original three, which had shouldered the entire traffic burden since 1896. Also GE products, this "new" power was quite different from the pioneers. These stubby boxcabs each had four motors and axles mounted in a rigid frame, and were specifically built for freight service. Generally they were run in semi-permanently coupled pairs, supplemented after 1906 by a single identical "motor" which could be used to form a three-unit set.

With the Waverly extension and third-rail system in place, the electric operation got yet another unique operational twist. East of York Road in Waverly, B & O built an interlocking tower and a center track to hold the electric locomotives between moves. To keep train movements fluid on the busy Belt Line, the eastbound trains did not stop to uncouple their "motor" helpers; the entire operation was done on the fly. Approaching York Road from the west, the train slowed to 8 mph while the steam locomotive began working to pick up the load. The electric then uncoupled at speed and raced ahead into the center "pen" track. The moment it was safely clear, the fast-moving

Photos at left and center: *During the first two decades of the 20th Century, camelbacks were the characteristic power on B & O trains between Philadelphia and Jersey City.* ***At left:*** *CNJ No. 595 brings an all-steel B & O train through Aldene, New Jersey in 1916. Smithsonian Institution Collection, C. B. Chaney photograph.*

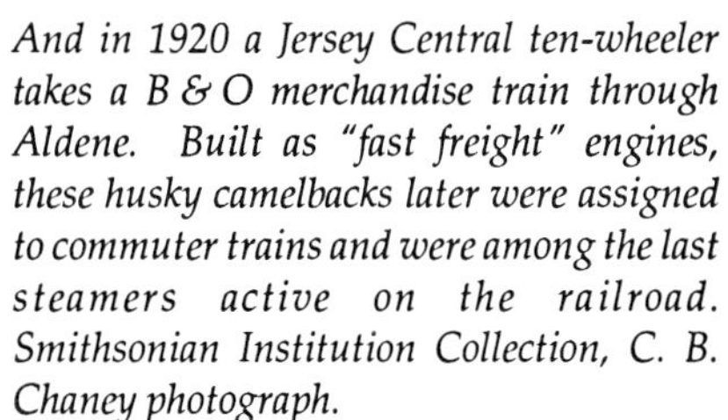

And in 1920 a Jersey Central ten-wheeler takes a B & O merchandise train through Aldene. Built as "fast freight" engines, these husky camelbacks later were assigned to commuter trains and were among the last steamers active on the railroad. Smithsonian Institution Collection, C. B. Chaney photograph.

tower operator lined the switch for the main track, allowing the train to accelerate past. Then, if no westbound was close, he quickly set up the route for the light "motor" to reverse and return to Camden. This seemingly frantic system generally worked smoothly, routinely handling thirty or more trains a day during the peak years.

During his brief reign, however, Loree was unimpressed with the entire concept of B & O's Baltimore Belt Line and looked for ways to supplement or replace it completely. To his efficiency-conscious mind, the grades, curves, and short one-way electric operation created unnecessary cost and congestion. Together with the Pennsy's A. J. Cassatt, Loree attempted to revive the idea of an elevated line around the harbor — this time as a joint route with the Pennsylvania, which of course had even worse tunnel problems. Never begun, the Loree-Cassatt project simply became the first of a succession of proposals made over the next twenty years to bypass the two railroads' Baltimore bottlenecks. These took several forms, ranging from a more-or-less simple pooling of the two existing rail lines through the city to a grandly ambitious "outer belt" line" which would circle Baltimore some distance to the west and north, rejoining the B & O and PRR main lines as far east as Havre de Grace, Maryland. But none material-

A trip of the stubby little 1903-1906 freight "motors" takes an eastbound train upgrade near Sisson Street in Baltimore. Smithsonian Institution Collection, B & O photograph.

The famous flying cutoff in a 1943 photograph. The camera looks west from Waverly tower toward York Road as a pair of motors race into the center "pen" track. On the curve behind, the eastbound freight is picking up speed. B & O Historical Society Collection, B & O photograph.

ized, and both 19th Century routes remain today. Reduced traffic and modern motive power (diesels on the B & O line and electrics on the former PRR) have helped make them more livable, but not any more loved.

Loree did make one move vital to the Royal Blue Line's future security. Between 1902 and 1903, with the Pennsylvania's active support, the B & O and New York Central jointly acquired working control of the Reading. Each company bought slightly less than twenty percent of the Reading's voting stock. With the Reading also came control of the Jersey Central, which the Reading had picked up in 1901, effectively putting the two interlocking systems into one family.

Behind the Reading purchase was the "community of interest" program created by Cassatt and supported by J. P. Morgan. A basically sound railroad rooted in the eastern Pennsylvania anthracite coal fields, the unfortunate Reading had been the victim of two over-ambitious late 19th Century empire builders: Franklin Gowen, its president intermittently between 1869 and 1886, and Archibald A. McLeod, whose short but spectacularly disastrous reign ran from 1890 to 1893. Their expansion attempts produced three receiverships within thirteen years, and both Morgan and Cassatt were concerned about the Reading's stability and the possibility that some hostile outsider would use it to invade their territories. In addition, both the B & O and New York Central depended on the Reading-Jersey Central lines as their eastern outlets: the B & O for its New York services and the Central for access to the Philadelphia, eastern Pennsylvania, and New Jersey markets. The resulting B & O-NYC "community of interest" control helped solve everyone's problems and did indeed create stability

Union Station soon after opening in 1907. H. H. Harwood, Jr. Collection.

Pennsylvania Station in all its imperial spendor about 1912. Smithsonian Institution Collection, C. B. Chaney photograph.

and prosperity for the Reading. And although the Reading and CNJ both remained separately managed, B & O now had a clear (and increasingly louder) voice in their policies.

Two other Loree-PRR "community of interest" projects vastly improved the B & O's unhappy situation at Washington. Since 1873 the Pennsy had heavily dominated the north-south freight business through the Washington gateway. It controlled the Potomac River crossing, and through its line to the Richmond, Fredericksburg & Potomac at Quantico, Virginia, it essentially monopolized traffic with the Atlantic Coast Line and the new Seaboard Air Line Railway systems. And although B & O limped along with its old Alexandria freight carfloat, the Pennsy directly connected there with the Southern Railway (the Virginia Midland's successor) and the Chesapeake & Ohio. At Washington itself, passenger terminals were a problem for both railroads, but especially the B & O. Its 1851 station at New Jersey Avenue had been substantially expanded in 1889, but remained cramped and crowded. In addition, all the southern passenger trains used the Pennsy's 6th and B Streets station six blocks away, further hampering B & O's ability to solicit through business. But both B & O and Pennsy were under extreme civic pressure to replace their stations and approach tracks, which were considered hindrances and eyesores. The Pennsylvania's approach tracks arrogantly occupied the Mall at 6th Street.

Fortunately the expansive Cassatt saw the necessity for high-capacity joint passenger and freight facilities in the area, and Loree was a willing partner. Planning began in 1901 for a vast new freight yard at Alexandria, which B & O would use along with the PRR, RF & P, Southern, and C & O. Potomac Yard, as it was named, opened in August 1906, and — thanks to some PRR trackage rights — B & O once again became an equal competitor for northeast-southeast freight traffic, including the ever-growing flow of fruits and vegetables into the Baltimore, Philadelphia, and New York markets.

The passenger terminal took slightly more time but produced one of Washington's most revered landmarks, Union Station. Work began in 1903, jointly financed and controlled by the Pennsy and B & O; on October 27, 1907 Union Station opened with the arrival of B & O train No. 10 from Pittsburgh. Afterwards B & O had no worries over station capacity or an appropriately grandiose entrance to the nation's capital. And since the PRR and all southern lines used the station, B & O passengers to or from the South had an easy transfer. But the Pennsy kept its tight hold on the through southern train services; except for a single short-lived through parlor car run between New York and Richmond (from September 1908 to May 1912), B & O never attempted a challenge.

In all, the early 1900s produced a physically strong, faster, corporately cohesive Royal Blue Line. Although it could not approach the Pennsy in total traffic volume, at least it could equal the PRR's running time and surpass it in advertising creativity. But Pennsy's A. J. Cassatt hardly had been idle either. Under him the railroad was undergoing the greatest rebuilding in its imperial history.

One small sample on the PRR's New York-Washington route was a track revision and new station at West Philadelphia, completed in 1903. Located at 32nd and Market Streets, the new West Philadelphia station became the main Philadelphia station for all New York-Washington runs, eliminating the time-consuming backup move to Broad Street Station. Certainly West Philadelphia was less conveniently located than Broad Street. In fact it was even farther from the city's center than B & O's station, but it was also served by Pennsy suburban trains and several streetcar lines on Market Street. After 1907 it also was reached by the new Market Street subway-elevated line. Whatever the drawbacks, the Pennsy felt that the time savings compensated. In 1906 it further accelerated its Washington schedules by eliminating its traditional Philadelphia engine changes; two years later its New York-Washington train crews also started working through. B & O could do neither of these. Royal Blue Line trains continued to exchange B & O, Reading, or CNJ engines and crews at Philadelphia, and would do so until 1926.

But Philadelphia was only a minor sideshow for Cassatt. Simultaneously he was fashioning his crown jewel, a direct entrance to Manhattan with a massive midtown passenger terminal. Since 1891 Pennsy engineers had been studying how to get across the Hudson into the city, finally settling on electrified tunnels. Clear early warning signs for the B & O came in 1900 when Cassatt bought control of the Long Island Rail Road. Eventually, LIRR property was to be used to store and service passenger cars for the new New York station and to connect with the New Haven's main line into southern New England. The full New York terminal project was officially announced in December 1901, although much remained to be done before any construction could start.

Even for the Pennsy it was an enormous undertaking. Thirteen miles of fully electrified new railroad would leave the main line east of Newark, New Jersey, tunnel under the Hudson as well as the full width of midtown Manhattan and the East River, and end in a large support yard and turning loops located in Queens on Long Island. And then there was the station itself: a huge neoclassical structure partly modeled on the Roman Baths of Caracalla, occupying two full blocks. Its site at 32nd Street between 7th and 8th Avenues was in a bit of an urban backwater, but was a block from Herald Square and close to what had become New York's commercial center. Ironically, the project's chief engineer was Samuel Rea, who had planned B & O's Baltimore Belt Line.

While the new Pennsylvania Station always would be the railroad's primary New York terminal, an adjunct of the project would offer Pennsy passengers a convenient way to reach certain other parts of the city. Almost coincidentally, work was started on the independently-owned Hudson & Manhattan Railroad, a New York-New Jersey rapid transit line. The H & M, familiarly called the "Hudson Tubes" (and now PATH), was planned to connect lower and midtown Manhattan with Jersey City and three New Jersey railroad terminals: those of the Pennsy, the Erie, and the Lackawanna. The H & M also intended to extend to Newark over the PRR's right of way.

The Pennsylvania and the H&M services were to be tied together at a point called Manhattan Transfer, an isolated spot in the "Jersey Meadows" east of Newark where all of the Pennsy's Penn Station trains would stop to exchange steam and electric locomotives. Here elevated platforms were built to allow PRR passengers to transfer to adjacent "Tubes" trains which took them to lower Manhattan (Hudson Terminal at Cortlandt Street, now the World Trade Center site) and to various stations along lower 6th Avenue and at Herald Square.

Construction work on the Penn Station project started in mid-1903 and took seven years; the first Pennsy train entered the new terminal November 27, 1910. (Cassatt himself never lived to see his completed monument; he died in 1906.) Partial Hudson & Manhattan operations began in 1908, and by 1911 the "Tubes" trains were directly connecting with all PRR passenger schedules at Manhattan Transfer.

The effect on the B & O obviously was devastating. Pennsylvania passengers now were whisked directly to midtown New York City with no train change or ferry trip. Admittedly, Penn Station was mildly inconvenient in its early years; the adjacent 7th Avenue subway was not opened until 1918 and initially the nearest first-class hotels were a block or more away. But few cared when the alternative was a ferry trip to lower Manhattan or 23rd Street. And for passengers bound for the downtown financial district, it was a simple matter to walk across the platform at Manhattan Transfer and into the waiting Hudson Tubes train bound for Hudson Terminal.

One of B & O's original Pacifics, No. 2107, takes a light five-car New York train out of Washington Union Station about 1910. Smithsonian Institution Collection, C. B. Chaney photograph.

As one of its by-products, the Penn Station project gave the Pennsy another marketing advantage. The extensive tunnels forced the railroad to equip all its New York trains entirely with all-steel cars. Actually it had introduced its first all-steel coach in 1906, and in 1907 it had begun producing its prolific standardized (albeit Spartan) P-70 class steel coaches. B & O had none at the time Penn Station opened; its first forty all-steel coaches did not arrive until 1911. In the meantime it could only advertise its latest "steel underframe luxury coaches" — ten blue-painted wooden-body A-14 class coaches turned out in 1910 and assigned to such premier trains as the *Royal Limited* and *Royal Special*.

Although marooned at Jersey City, B & O's new president, Daniel Willard, was hardly ready to give up. But for the moment he had to make the best of it. The resources of the B & O, Reading, and CNJ were certainly not up to duplicating the Pennsy's feat, which had cost almost $113 million. In the years afterwards, all of the "New Jersey railroads" (which besides the B & O/Reading/CNJ group included the Erie, Lehigh Valley, Lackawanna, NYC's West Shore line, and its tenant, the New York, Ontario & Western) were involved in various proposals to build some joint Manhattan terminal, but nothing was ever done. Nothing materialized either on a plan to build a Hudson & Manhattan "Tubes" branch to the CNJ's Jersey City terminal, and the Jersey Central and its partners always remained completely dependent on the ferries.

Improvements continued on the Royal Blue Line, although without the dramatic flair of earlier years. By 1910 Willard had a high-capacity Washington-Philadelphia line; in fact it would not require substantial rebuilding again for at least eighty years. B & O's original light P-class Pacifics were assigned to some Philadelphia runs, and in 1910 the railroad received 26 new heavy Atlantics, many of which were put to work on the route. These Baldwin-built A-3s had 80-inch drivers, the largest yet used on B & O, and were its heaviest 4-4-2s. But as steel passenger cars finally began to arrive, the A-3s were soon supplanted by heavier Pacifics, particularly the P-3s of 1913. Three years later Reading's own shops began turning out its first Pacifics, the lithe G-1s, to handle B & O's trains north of Philadelphia as well as the Reading's own New York and Atlantic City passenger business. Designed in the Reading's tradition of fast power, the exceptional G-1s had 80-inch drivers and, despite their off-center headlights and wide Wootten fireboxes, they were some of the handsomest power on the "Blue Line."

In this period too the Baltimore-Philadelphia line played host to what was unquestionably its oddest and most powerful locomotive, no less than a 2-8-8-8-2 triplex compound articulated. In 1914 the Baldwin Works turned out the first of three such monsters for the Erie Railroad, Erie's No. 5014, the *Matt H. Shay*. Before delivering the 426-ton machine to its new owner, Baldwin arranged with both the B & O and Reading to test it on their lines serving the Eddystone plant. The perfect antithesis of anything suitable or needed on the Royal Blue Line, the *Shay* nonetheless made several trips to Bay View in Baltimore.

The Baltimore Belt Line also began a slow process of replacing its early and now-primitive electric locomotives with power able to handle the heavy steel passenger trains and lengthening freights. In 1910 B & O settled on a standardized double-truck four-motor 1100-hp GE "steeple cab" design based on the Michigan Central's recent Detroit River Tunnel electrics. Between 1910 and 1927 it bought a total of eight new "motors," always ordering two at a time. Although they often operated singly in the earlier years, two-unit sets gradually became the Belt Line's normal motive power for passenger trains and most freights; heavier freights got four units "doubleheaded" with two engine crews. The pioneering 1895 motors were retired by 1916 and the last of the odd little 1903-1906 freight boxcabs were taken out of service in 1929.

By 1916 B & O could advertise that all its through New York-Washington schedules were "electric lighted, vestibuled steel trains." But despite the husky Pacifics, overall schedules were little different from ten years earlier. Eight through schedules operated each way, with five hours still the fastest terminal-to-terminal (including ferry) running time. And as before, only two premier trains each way made the trip in five hours: the *Royal Limited* (which by then was carrying coaches too) and the *Royal Special*, a name which first appeared May 29, 1910. (By the evidence of period photos, the *Royal Special* usually consisted of only three cars, which included a coach, diner, and parlor-observation car.) Aside from these, the fastest Royal Blue Line train took five hours 22 minutes — slightly slower, in fact, than in 1904.

At the end of 1916 too, the "Royal Blue" image itself suddenly vanished, partly a victim of the war and changing attitudes, and partly from the influx of newer cars painted in what had become the railroad industry's standard scheme of olive green. In December 1916 the *Royal Limited* was renamed the *National Limited*, and the *Royal Special* briefly became the *Capitol Special* and then the *National Special*. All references to the Royal Blue Line in B & O timetables and advertising ceased, and the New York trains became, well, just New York trains. But if officially abandoned by the railroad, the Royal Blue name lingered in the minds of both passengers and railroaders, including Daniel Willard, who revived it two decades later. And even today, Conrail refers to certain former Reading facilities in Philadelphia as the "Blue Line bridge" or the "Blue Line connection."

Another Pacific with another New York express passes the Washington Terminal Company's engine facility at Ivy City. Smithsonian Institution Collection, C. B. Chaney photograph.

Meanwhile the Pennsy generally had tightened its times. Now rolling behind the chunky but fast E-6 class Atlantics, several PRR trains covered the distance in about five hours and fifteen minutes, although the *Congressional Limited's* best time remained at five hours. Now, however, B & O and PRR running times were no longer comparable; Pennsy's were measured from Penn Station in the center of New York and B & O's from the lower Manhattan ferry terminal at Liberty Street.

B & O's advertising department desperately looked for something good to say about Jersey City and the ferries. It was clearly a losing battle, but the railroad certainly tried: "Practically a Union Station at New York City" said a 1916 ad, which proceeded to point out that B & O's Jersey City terminal also served Reading and Jersey Central trains, while the CNJ ferries were convenient to ferries of the Erie, the Lackawanna, and the NYC's West Shore line. Sadly, few passengers were heading for these other railroads, and the ad somehow neglected to mention that PRR passengers using the Hudson Tubes could reach the Erie and Lackawanna terminals quicker and more directly anyway. Somewhat wistfully the ad concluded: "The most magnificent view of New York City is that of its skyline, and it cannot be seen in any other way than by the ferry ride from Jersey City." That was indeed true, especially at dusk, and remained so as long as the ferries ran. But few passengers seemed to care.

For those passengers who noticed, however, the Jersey City terminal was significantly modernized. By 1911 the 1889 station and twelve-track trainshed were handling 14.5 million passengers a year — most of them hurrying to and from the Jersey Central's own suburban trains and, in summer, to the New Jersey seashore. Starting in 1912 the CNJ dismantled its old trainshed, expanded the terminal track layout from twelve to twenty tracks, built a modern concrete and steel Bush-design umbrella trainshed, and rebuilt the station's train and ferry concourses. Carried on without interrupting the 370 trains which used the station every day, the project was finished in the summer of 1914. At the same time, the CNJ built a mammoth new engine terminal with twin roundhouses near Communipaw Avenue, capable of servicing 300 locomotives a day.

While Penn Station seriously dampened B & O's New York passenger prospects, its freight outlook was at least comparatively cheerier. Although B & O could not directly serve industries on the CNJ and Reading lines, it often included itself in their shipment routings west of Philadelphia. In addition, the once-isolated Staten Island shore line began to develop. In 1905 Procter & Gamble opened the first section of its Port Ivory soap plant, located near the SIRT's Arlington yard. Over the years Port Ivory grew into a major on-line SIRT/B & O shipper and receiver. Also beginning in 1905, the older American Dock & Trust Company's pier and warehouse complex at Tompkinsville (just east of St. George) was significantly expanded and rail sidings installed. Later, various other piers and warehouses were gradually established along the island's northeastern shore in the Tompkinsville-Stapleton area.

In addition, bituminous coal continued flowing through the St. George docks in increasing volumes. In 1905 a steam-powered McMyler car dumper was added to the pier complex, greatly speeding the flow of coal between cars and vessels. The McMyler machine hoisted the hopper individually and rotated each, dumping its contents out the car's top into a chute leading to the vessel's hold.

In fact, by 1912 the relatively cramped St. George yard was becoming overcrowded with the growing coal and general freight business. By then the marine facilities included six open piers, two covered piers, the three coal piers, and three "float bridges" used to transfer cars to and from the carfloats. Five yard crews worked there on all shifts, shuffling cars off and on the piers and carfloats and trying to keep the yard fluid. Two other 24-hour crew

Physically separate and a galaxy apart from the Royal Blue Line's glamor, St. George terminal did the drudge work that earned most of the money. Obviously it was at capacity in this 1912 scene. B & O Historical Society Collection, B & O photograph.

Freight cars were transferred to and from carfloats at these float bridges at St. George. B & O Historical Society Collection, B & O photograph.

relays worked the support yard at Arlington, east of the Arthur Kill drawbridge, and an average of four day and night crews handled Staten Island road freight trains — most of which were powered by 0-6-0 or 2-8-0 camelbacks, many of them owned by B & O. And although B & O hardly wanted them, the SIRT's little camelback 4-4-0s and Forney-style steamers were wheeling no less than 334 trains a day in 1912. The year earlier the line had carried 6.2 million passengers. All that was no bonanza, however; the Staten Island line's maximum fare was thirty cents, and its average passenger rode less than five miles and paid five cents.

The SIRT, incidentally, was never used as a main line passenger link except for special movements. Although the B & O had vague early plans to route its New York passenger trains through Staten Island and roll them onto a carferry for New York, it never pursued the idea. But according to company records, from 1900 to 1903 the B & O and Jersey Central did attempt an unsuccessful joint through connecting service between St. George and the CNJ station at Plainfield, New Jersey. Plainfield, 24 miles west of Jersey City, was a regular stop for B & O's New York trains, and two round trips a day were run to connect with the B & O here. *Official Guides* of this period do not show these trains and nothing more seems to be known about this mysterious and intriguing operation.

Along the Manhattan and Brooklyn waterfronts the B & O slowly developed a collection of leased pier stations, most of them used for perishable and merchandise carloads floated to and from St. George. About 1898 it also became a joint user of the Harlem Transfer Company's amazingly compact on-land rail freight terminal in the Bronx at 135th Street west of Third Avenue. And beginning in the early 1900s an extensive carfloat business developed between B & O and a group of newly-built switching railroads serving piers and warehouses along the Brooklyn shore — notably the New York Dock Company, the Brooklyn Eastern District Terminal, Bush Terminal, and Jay Street Terminal (which after 1909 was called Jay Street Connecting Railroad). Although operated by independent companies

Tugs like the 1915 George M. Shriver moved carfloats and lighters between St. George and points around the harbor. B & O Museum Collection.

Coal, B & O's major inbound commodity at New York, was transferred to harbor barges by this McMyler car dumper, also shown in 1912. B & O Historical Society Collection, B & O photograph.

Opened in 1914, B & O's "inland" New York warehouse was an active freight facility into the 1970s. It still stands, although no longer railroad-owned or rail-served. B & O Museum Collection, B & O photograph.

Photos left and below: *At Tottenville, SIRT trains connected with the railroad's picturesque Perth Amboy ferry. Both photographs date to 1912. B & O Historical Society Collection, B & O photographs.*

serving all New York-area railroads, these terminals were considered B & O stations when handling any B & O-routed shipments. Finally, early in 1914 B & O completed a large new nine-story concrete warehouse at its key 26th Street Manhattan freight terminal. The 26th Street facility was subsequently expanded to include less-than-carload freight sheds, and remained B & O's most important single New York terminal through the 1960s.

As B & O's New York business expanded, so did its "navy" there. By 1916 no less than ten steam-powered tugs were operating in the harbor, ranging from the 1865 *Narragansett* to the newly-delivered (in 1915) *George M. Shriver*. B & O tugs were distinguished by three orange bands on their funnels, a marking which remained the railroad's harbor insignia until diesel tugs appeared in 1952 and 1953. And although the Manhattan ferries were no longer railroad-operated, the Staten Island line still owned two boats which shuttled between Tottenville and Perth Amboy at the other end of the island. This service too was long-lived; the railway ran it until 1948 and the service itself survived until 1963.

The United States's entry into World War I created a nightmare for the Staten Island terminals and everywhere else in New York Harbor. But it also brought a surprise and certainly an undreamed-of blessing: on April 28, 1918, Penn Station became B & O's New York passenger terminal.

PENN STATION AND AFTERWARDS 1918-1930

Although the use of Penn Station put the B & O back into genuine competition, the reasons behind it were less than glorious. As the railroads approached paralysis under World War I traffic loads, the federal government felt that it was forced to take temporary control of all the major carriers. On December 28, 1917 the hastily formed United States Railroad Administration did just that, and immediately began working to unclog the traffic flows by operating the multitude of competing railroad lines as a single system.

Whereas freight movement was the USRA's most critical concern, certain passenger flows also needed attention — notably the New York-Washington route. Stated simply, while the Pennsylvania was staggering under the demands of the Washington business and troop movements, the B & O's trains were underused. By moving B & O services into Penn Station in April 1918, USRA Director General William Gibbs McAdoo hoped to take some pressure off the Pennsy and spread the loads a little more evenly. (The Lehigh Valley, which also had used CNJ's Jersey City terminal, was moved into Penn Station in September.)

Significant operating changes were needed to make the move. Since there was no immediately usable connection between the Jersey Central and PRR's Penn Station line, a new routing had to be worked out. (The currently used commuter train connection at Aldene, New Jersey was not built until 1967.) The solution was to use the traditional Reading route from Philadelphia to Manville, New Jersey, about two miles south of the Jersey Central at Bound Brook. At Manville the Lehigh Valley's double track New York main line crossed the Reading at grade with a direct track connection. The LV in turn crossed the PRR on the south side of Newark at a spot called West Newark Junction, now known as Hunter Tower. B & O's Penn Station trains thus would completely bypass the Jersey Central, using the Lehigh Valley to reach Pennsy trackage at Newark.

With this decided, new arrangements were needed to replace the old B & O-Reading-CNJ traffic agreement dat-

This less-than-perfect shot is nonetheless a great rarity; it is one of the very few photographs taken during B & O's operation into Penn Station. Here, Reading G-1 Pacific 108 takes B & O train 507 west on Lehigh Valley tracks at Aldene, New Jersey in 1920. The bridge behind crosses the Jersey Central's main line. Smithsonian Institution Collection, C. B. Chaney photograph.

From 1913 to 1919 or so, P-3 Pacifics like the 5120 handled the Washington-Philadelphia leg of the New York passenger runs. New York-Washington train 501 passes Halethorpe, Maryland in 1919. Smithsonian Institution Collection, C. B. Chaney photograph.

ing (with modifications) back to 1886. Under that agreement the three railroads were each responsible for their own part of the haul and received a division of the revenue. But for the new service, a form of trackage rights contract was used; between Philadelphia and Penn Station, B & O had operating trackage rights over the Reading, Lehigh Valley, and Pennsylvania. Over this section, however, B & O trains were hauled by Reading locomotives (usually G-1 Pacifics) and manned by Reading crews. At Manhattan Transfer, of course, the Reading power was exchanged for the Pennsy's peculiar but competent DD-1 class third-rail electrics which handled all trains into Penn Station. In common with the Pennsy's passenger trains, B & O equipment was stored and serviced at Sunnyside yard on Long Island. B & O's New York freight was unaffected by the USRA order and continued moving over the old B & O-Reading-CNJ-SIRT route.

The first B & O Penn Station services essentially were USRA expediencies to add capacity to the New York-Washington route. B & O's traditional eight New York round trips were cut to six each way and made into strictly New York-Washington runs; all through B & O cars for points west of Washington were eliminated. They were slowed too; the five-hour schedules disappeared and the fastest runs over the B & O route took five hours 35 minutes. By November 1918 one westbound B & O train was carrying Cincinnati sleepers again and, oddly, one B & O eastbound brought Southern Railway and Chesapeake & Ohio sleepers from Washington to New York. By July 1919 B & O's New York services began to look more like they did before the emergency. The number of trains was up to seven each way and most of the traditional through B & O western

Succeeding the P-3s were the P-5s, designed and ordered by the USRA in 1919. Four years after delivery, P-5 No. 5204 storms south on the Washington Branch with train 505 from New York. Smithsonian Institution Collection, C. B. Chaney photograph.

sleeper runs had been restored, but five and one half hours was still the best running time. Schedules were speeded up again in the early 1920s, but the total service remained at seven trains each way, none of which were as fast as five hours.

In 1919 the USRA assigned to the B & O thirty Baldwin- and Alco-built standard-design light Pacifics, many of which were put on the Washington-Philadelphia portion of the New York runs. (Reading Pacifics continued hauling B & O trains east of Philadelphia.) Handsome but unglamorous, in keeping with the Royal Blue Line's new image, these utilitarian P-5 class engines with their smaller 73-inch drivers became the backbone of the Washington-Philadelphia operations through much of the 1920s.

Federal control of the railroads formally ended March 1, 1920; and with it, B & O's right to use Penn Station technically also ended. Somehow, however, the diplomatic Daniel Willard managed to negotiate a new agreement with the PRR, extending the operation another five years. B & O continued to be limited to seven trains each way with no more than twelve cars in a train. And as usual, B & O was not permitted to handle any through cars for points south of Washington or north of New York. That restriction hardly mattered; except for the earlier Richmond parlor car and the USRA emergency routings, B & O had always stayed out of the Pennsy's domain.

The new Penn Station contract was signed July 13, 1921, and afterwards Willard was optimistic enough to order his engineering department to study a new "permanent" alternative route between the Reading and the Pennsy in New Jersey. Apparently there were some problems in using the Lehigh Valley segment between Manville and Newark. In particular, Willard was interested in using the Reading's Port Reading branch from Manville (adjacent to the LV connection) to the PRR main line at Menlo Park, New Jersey, east of Metuchen. The B & O engineers also considered rebuilding a section of the SIRT between the Jersey Central at Cranford and the Pennsy near Linden for use as a passenger connection.

The optimism wilted quickly. On February 1, 1922 Pennsylvania president Samuel Rea wrote Willard to tell him that when B & O's five-year Penn Station agreement expired in 1925, it would not be renewed. Penn Station was simply too congested, Rea said, to accommodate the B & O too. As the September 25, 1925 eviction date approached, Willard dickered for more time and got Rea to agree to a one-year extension. But afterwards Rea's harder-nosed successor, W. W. Atterbury (who became president in October 1925) refused any further extensions. B & O had to be out of Penn Station by September 1, 1926, or sooner, if it could.

Willard's only practical alternative was to return to the Jersey Central's Jersey City terminal and its ferries — or to admit competitive defeat and pull his passenger services back to Philadelphia or Baltimore. A mixture of pride and optimism made him stick to Jersey City, "temporarily," it was announced. In 1926 all large railroads were looking forward to greater traffic growth and prosperity; Willard, for one, felt that eventually the various other New Jersey-based railroads somehow would unite and build a Manhattan terminal of their own. In the meantime he not only would keep B & O's flag in New York, but would spare nothing to make potential passengers forget Penn Station and that other railroad.

His methods were expensive indeed, but creative and exciting. Willard's strategy had three principal parts. He would match Pennsy running times as closely as practical (and, for the first time, run locomotives and crews through Philadelphia without change). Also, he would take advan-

Pennsylvania Station
New York City
7th Avenue and 32nd Street

BALTIMORE AND OHIO RAILROAD
SCHEDULE OF TRAINS BETWEEN
New York, Philadelphia, Baltimore and Washington

Union Station
Washington
Massachusetts and Delaware Aves.

WESTWARD

Baltimore and Ohio Trains arrive at and depart from New York on Eastern Standard Time.

Miles	STATIONS In Effect April 30, 1922	511 Daily	21 Daily Note	5 Daily	⊙1 Daily	★✣7 Daily	525 Daily	9 Daily	3 Daily	17 Daily
		AM	AM	AM	AM	PM	PM	PM	PM	PM
0.0	Lv **New York, N. Y.** (Pennsylvania Station)	1.00	....	7.50	10.50	12.50	2.40	4.45	5.50	....
0.0	Lv **New York** (Hudson Terminal)	1.00	....	7.48	10.50	12.50	2.30	4.42	5.50	....
1.0	Lv Jersey City, N. J. (Exchange Place)	1.03	....	7.51	10.53	12.53	2.33	4.45	5.51	....
8.8	Ar ●Manhattan Transfer	1.14	....	8.04	11.04	1.04	2.54	4.59	6.05	....
0.0	Lv Newark (Park Place)	1.00	....	d 7.53	10.50	12.50	2.40	4.52	5.50	....
1.3	Ar ●Manhattan Transfer	1.03	....	7.56	10.53	12.53	2.43	4.55	5.53	....
8.8	Lv ●Manhattan Transfer	1.18	....	8.08	11.08	1.08	2.58	5.03	6.09	....
34.1	Lv Bound Brook (Lehigh Valley)	....	....	a 8.45	a 11.45	a 1.45	a........	....	a........	....
86.9	Lv Wayne Junction, Pa.	3.16	7.45	9.49	12.49	2.49	4.37	6.42	8.01	....
94.7	Ar **Philadelphia** (24th and Chestnut)	3.34	8.00	10.05	1.05	3.05	4.53	6.58	8.15	....
94.7	Lv **Philadelphia** (24th and Chestnut)	3.44	8.02	10.10	1.10	3.10	4.57	7.03	8.20	8.35
106.6	Lv Chester	....	8.21	10.29	1.29	3.29	5.16	7.22	h 8.38	8.54
119.7	Lv Wilmington, Del.	4.27	8.37	10.47	1.47	3.47	5.34	7.40	8.56	9.13
131.9	Lv Newark, Del.	4.48	8.54	....	2.03	....	5.50	....	....	9.30
189.1	Ar **Baltimore, Md.** (Mt. Royal Station)	6.20	10.05	12.10	3.10	5.12	6.57	9.03	10.20	10.42
190.6	Ar **Baltimore** (Camden Station)	6.25	10.10	12.15	3.15	5.17	7.02	9.08	10.25	10.47
190.6	Lv **Baltimore** (Camden Station)	6.35	10.13	12.18	3.20	5.21	7.04	9.12	10.30	10.50
227.4	Ar **Washington, D. C.** (Union Station)	7.30 AM	11.05 AM	1.10 PM	4.15 PM	6.15 PM	7.50 PM	10.00 PM	11.25 PM	11.45 PM
	(Connections for Richmond, Va.)									
.0	Lv **Washington** (R. F. & P.)	8.55	11.45	3.15	5.10	7.45	9.40	3.15	3.15	3.15
116.5	Ar **Richmond, Va.**	n12.20 PM	b 3.00 PM	b 6.25 PM	b 9.15 PM	m11.05 PM	m11.05 PM	b 7.10 AM	b 7.10 AM	b 7.10 AM

CAR SERVICE

511—"WASHINGTON EXPRESS." Drawing-room Sleeping Cars and Coaches New York to Baltimore and Washington (cars ready for occupancy Pennsylvania Station 9.00 p. m. Standard Time;) Philadelphia to Washington (open for occupancy 9.00 p. m. Standard Time). Passengers may remain in Sleepers at Baltimore until 6.30 a. m. and Washington until 7.30 a. m. Standard Time.

21—"THE WASHINGTONIAN." Parlor Car, Parlor Dining Car and Coaches Philadelphia to Washington.

5—"NEW YORK-CHICAGO LIMITED." Coaches, Drawing-room Parlor Cars and Dining Car New York to Washington.

1—"NEW YORK-CINCINNATI-ST. LOUIS LIMITED." Coaches and Drawing-room Parlor Cars New York to Washington; Dining Car New York to Baltimore.

7—"NEW YORK-CHICAGO SPECIAL." Coaches and Drawing-room Parlor Cars New York to Washington; Dining Car New York to Philadelphia.

525—"WASHINGTON EXPRESS." Coaches and Drawing-room Parlor Cars New York to Washington; Dining Car Philadelphia to Washington.

9—"NEW YORK-CHICAGO EXPRESS." Coaches, Drawing-room Parlor Cars, Club Car and Dining Car New York to Washington.

3—"NEW YORK-CINCINNATI-ST. LOUIS EXPRESS." Coaches, Drawing-room Parlor Car and Dining Car New York to Washington.

17—"PITTSBURGH EXPRESS." Coaches Philadelphia to Pittsburgh; Drawing-room Sleeping Cars Philadelphia and Washington to Pittsburgh.

• Daily. † Daily except Sunday. ● For transfer of passengers only. a Stops to receive passengers for stations west of Philadelphia. b Broad St. Station—Local trains and through trains of R. F. & P. in connection with Atlantic Coast Line arrive at and depart from Broad St. Station. d Sunday 7.58 a. m. f Stops on signal to receive or discharge passengers. h Stops to receive passengers for stations west of Cumberland. m Main St. Station—Through trains of R. F. & P. in connection with Seaboard Air Line arrive at and depart from Main St. Station. Sleepers for Baltimore and Washington open for passengers in Pennsylvania Station at 10.00 p. m. Passengers may remain in Sleepers at Baltimore until 6.30 a. m..and Washington until 7.30 a. m Standard Time.
★ Connects with Old Bay Line and Chesapeake Line Steamers leaving Baltimore daily, except Sunday, at 6.30 p. m. for Old Point Comfort and Norfolk. On Sundays service between these points is operated alternately between the Chesapeake Line and the Old Bay Line. Tickets of either line will be honored on Sundays. Street cars direct from Camden Station to the wharf.
✣ Connects with York River Line Steamers leaving Baltimore except Sundays at 6 p. m. for West Point and Richmond. Street car direct from Camden Station to the wharf.
⊙ Connects with Norfolk and Washington Steamers leaving Washington daily (7th Street Wharf) 6.30 p. m. for Old Point Comfort, Norfolk and Portsmouth.
Annapolis Short Line trains, formerly arriving at and departing from Camden Station, now arrive at and depart from W. B. & A. Terminal, Lombard and Howard Streets, two blocks from Camden Station. Direct street car service.
NOTE—Train No. 21 receives connection at Wayne Junction from Syracuse, Buffalo, Rochester, Ithaca, Binghampton, Sayre and Scranton

EASTWARD

Miles	STATIONS In Effect April 30, 1922	528 Daily	12 Daily	4 Daily	8 Daily	524 Daily	2 Daily	22 Daily	6 Daily	20 Daily
	(Connections from Richmond)	PM		AM	AM	AM	AM	PM		PM
.0	Lv **Richmond, Va.** (R. F. & P.)	b 8.15	....	....	b 5.25	b 9.15	b 11.55	....	....	† 2.00
116.5	Ar **Washington, D. C.**	11.50	....	....	8.35	12.25	2.35	....	....	6.15
		NT	AM	AM	AM	PM	PM	PM	PM	PM
0.0	Lv **Washington, D. C.** (Union Station)	12.25	2.20	7.15	9.25	1.00	3.10	4.00	5.00	7.00
36.8	Ar **Baltimore, Md.** (Camden Station)	1.20	3.13	8.03	10.13	1.48	3.58	4.50	5.50	7.50
36.8	Lv **Baltimore** (Camden Station)	1.25	3.20	8.07	10.18	1.51	4.02	4.53	5.55	8.05
38.3	Lv **Baltimore** (Mt. Royal Station)	1.35	....	8.12	10.23	1.56	4.07	4.58	6.00	8.10
95.5	Lv Newark, Del.	....	....	9.23	11.34	....	5.18	6.09	7.11	9.41
107.7	Lv Wilmington	3.18	4.58	9.40	11.51	3.22	5.35	6.25	7.28	10.00
120.8	Lv Chester, Pa.	....	g 5.17	9.57	12.08	3.39	5.52	6.42	7.45	10.20
132.7	Ar **Philadelphia** (24th and Chestnut)	3.58	5.38	10.17	12.27	3.57	6.10	7.00	8.04	10.40
132.7	Lv **Philadelphia** (24th and Chestnut)	4.04	5.43	10.22	12.32	4.02	6.15	....	8.09	....
140.5	Ar Wayne Junction	4.23	6.01	10.39	12.48	4.20	6.31	....	8.25	....
193.3	Ar Bound Brook, N. J. (Lehigh Valley)	n 5.47	....	n........	n........	n 5.24	n........	....	n 9.30	....
218.6	Ar ●Manhattan Transfer	6.31	7.50	12.21	2.29	6.01	8.12	....	10.11	....
0.0	Lv ●Manhattan Transfer	6.37	7.59	12.27	2.47	6.07	8.17	....	10.17	....
1.3	Ar Newark (Park Place)	6.40	8.02	12.30	2.50	6.00	8.20	....	10.20	....
218.6	Lv ●Manhattan Transfer	6.35	7.54	12.25	2.33	6.05	8.16	....	10.15	....
226.4	Ar Jersey City, N. J. (Exchange Place)	6.47	8.08	12.38	2.47	6.17	8.27	....	10.27	....
227.4	Ar **New York, N. Y.** (Hudson Terminal)	6.52	8.13	12.41	2.51	6.20	8.30	....	10.30	....
227.4	Ar **New York** (Pennsylvania Station)	6.50 AM	8.08 AM	12.40 PM	2.48 PM	6.20 PM	8.30 PM	....	10.30 PM	....

CAR SERVICE

528—"NEW YORK EXPRESS." Coaches and Drawing-room Sleeping Cars Washington to New York (ready Union Station 10.00 p. m. Standard Time), and Baltimore to New York (ready Camden and Mt. Royal Stations 9.00 p. m. Standard Time). During period daylight saving plan is in effect at New York City (April 30th to September 24th), passengers will be required to vacate Sleeping Cars of Train 528 at New York immediately on arrival.

12—"METROPOLITAN SPECIAL." Coaches and Drawing-room Sleeping Cars St. Louis, Cincinnati and Fairmont to New York; Pittsburgh to Philadelphia, Drawing room Sleeping Car Washington to Philadelphia (ready for occupancy 10.00 p. m. Standard Time). Passengers may remain in Sleepers at Philadelphia until 6.30 a. m. Standard Time.

4—"CINCINNATI-WASHINGTON-NEW YORK EXPRESS." Coaches, Drawing-room Parlor Cars and Dining Car Washington to New York.

8—"CHICAGO-WASHINGTON-NEW YORK SPECIAL." Coaches, Drawing-room Parlor Cars and Dining Car Washington to New York.

524—"NEW YORK EXPRESS." Coaches, Drawing-room Parlor Cars, Club Car and Dining Car Washington to New York.

2—"ST. LOUIS-CINCINNATI-NEW YORK LIMITED." Coaches and Drawing-room Parlor Cars Washington to New York; Dining Car Baltimore to New York.

6—"CHICAGO-NEW YORK EXPRESS." Coaches, Drawing-room Parlor Cars and Dining Car Washington to New York.

22—"THE PHILADELPHIAN." Parlor Car, Parlor Dining Car and Coaches Washington to Philadelphia.

• Daily. † Daily except Sunday. ● For transfer of passengers only. Light-face figures A. M. time. **Black-face figures P. M. time.**
b Broad St. Station—Local trains and through trains of R. F. & P. in connection with Atlantic Coast Line arrive at and depart from Broad St. Station. f Stops on signal to receive or discharge passengers. g Stops to discharge passengers from Cumberland and stations west. m Main St. Station—Through trains of R. F. & P. in connection with Seaboard Air Line arrive at and depart from Main St. Station. n Stops to discharge passengers from stations west of Philadelphia.
Sleeping Cars on Train No. 528 open for passengers in Washington (Union Station) 10.00 p. m. and Baltimore (Mt. Royal and Camden Stations) at 9.00 p. m. Standard Time. During period daylight saving plan is in effect at New York City (April 30th to September 24th), passengers will be required to vacate Sleeping Cars of Train 528 at New York immediately on arrival.
Annapolis Short Line trains, formerly arriving at and departing from Camden Station, now arrive at and depart from W. B. & A. Terminal, Lombard and Howard Streets, two blocks from Camden Station. Direct street car service.

B & O Penn Station services in 1920. B & O Museum Collection.

Daniel Willard ran the B & O for 31 years, from 1910 to 1941, and took particular proprietary pride in the old Royal Blue Line. Seen here at age 66 in 1927, he demonstrates why many B & O railroaders referred to him as "Uncle Dan." B & O Museum Collection.

tage of every technical innovation which promised greater speed or passenger comfort. And, most important, he would operate a direct trainside bus service connecting all B & O trains at Jersey City with a variety of points in New York City.

Underpinning the new operation was a trackage rights contract made with the Reading and Jersey Central in August 1926. Unlike the old Royal Blue Line arrangements, B & O was to have full control of its passenger train operations all the way to Jersey City. B & O power, cars, and crews would operate between Washington and Jersey City, and the Reading and Jersey Central were to be paid a rental fee based on train mileage. As in the past, however, the Jersey Central got extra reimbursement for its terminal expenses and for any special facilities the B & O needed. The Reading and CNJ also got revenue divisions for any local business B & O trains carried east of Philadelphia. Although the original trackage agreement specified B & O crews, in practice both B & O and CNJ crews were pooled for the 223.6-mile runs. Reading crews were added to the pool later, although all "foreign" railroad employees wore B & O uniforms.

Willard's next problem was motive power capable of making the through Washington-Jersey City runs at consistently high speeds. The P-5 Pacifics were adequate for the rather undemanding schedules of the Penn Station period, but Willard now had to move his trains as fast as he could to compensate for his New York shortcomings. By this time the Pennsylvania's immensely effective K-4 Pacifics were wheeling its Washington-New York trains at 80 mph and more, and B & O had to do the same.

His answer was a close copy of the K-4, to become B & O's legendary P-7 "President" class. Twenty of these heavy 80-inch-drivered Pacifics were ordered from Baldwin for 1927 delivery — coincidentally B & O's centennial year. Except for experimentals, they would be the heaviest, fastest, and most modern steam passenger power the railroad would ever own. Determined to make his "new" New York operation distinctive, Willard had his new P-7s painted olive green to match their trains, with understated

The Pennsy was moving its New York-Washington trains with brutishly heavy but fast power in the 1920s. Shortly before World War I, an E-6 Atlantic waits to take the Congressional Limited out of Manhattan Transfer. Smithsonian Institution Collection, C. B. Chaney photograph.

By the late 1920s the E-6s had been largely supplanted by the now-legendary K-4 Pacifics. In 1929 this one had ten cars in tow at Rahway, New Jersey. A. P. Formanek photograph.

gold and red striping. Most memorable, each carried the name of an American president — a deft touch to emphasize B & O's connection with Washington, DC as well as its own historical heritage. Royal Blue may have disappeared, but class had returned.

Most creative was Willard's solution to the New York terminal problem. By the mid-1920s the Pennsylvania had made itself into a mass hauler operating with Teutonic efficiency, "The Standard Railroad of the World" as it called itself with minimal modesty. Willard knew that he could not compete on those terms. Instead he would offer what the Pennsy usually did not: as close to an individualized personal service as possible. New York-bound passengers were met directly at their Jersey City train platforms by a fleet of deluxe motor buses, which then took them to a choice of major hotels and commercial centers in the city. Their baggage was checked, and was loaded and unloaded by porters. The New York-bound buses were driven onto the CNJ ferries for the ride across the river, then followed two routes: "uptown" buses rode the 23rd Street ferry, then drove to a terminal station at the Pershing Square Building on 42nd Street opposite Grand Central Terminal. Along the way they stopped at the Pennsylvania Hotel (opposite Penn Station, which was never mentioned), the McAlpin, the Waldorf-Astoria (then located at 33rd Street and 5th Avenue), and the Vanderbilt Hotel. The "downtown" route also ended at 42nd Street, but used the Liberty Street ferry and served the railroads' Consolidated Ticket Office at Broadway and Chambers Street, Union Square, and also the Vanderbilt Hotel.

Although the bus trip was short (fifty minutes at the most, counting the "stretching" time during the ferry crossing), the buses themselves were intercity "parlor" coaches, Yellow Coach Model Ys, fitted out with individual cushioned wicker seats and window curtains. Like the "President" Pacifics they were painted olive green, but with a blue belt strip, and were prominently lettered in gold "Baltimore & Ohio Train Connection". Although railroad-owned, they were operated by the Fifth Avenue Coach Company, New York's major bus line. (Later, Gray Line Motor Tours operated the B & O buses.) Incidentally, B & O *never* used the plebeian word "bus"; in ads, timetables, correspondence, and all other forms of communication, they were always more exalted "motor coaches."

To accommodate the trainside transfer at Jersey City, the Jersey Central assigned B & O the two most northerly platforms in the terminal. Tracks were removed from between the two platforms and paved over as a driveway so that the buses could load or discharge under cover directly opposite the railroad cars. To turn them within this narrow

Photos above and below: *Dan Willard's pride, the "President" P-7 class.* ***Above:*** *The portrait view of No. 5310 President Taylor at Washington in 1928 shows the engines' original appearance. L. W. Rice photograph.* ***Below:*** *By the time Bill Osborne shot the same engine scooping water on the Reading at Roelofs, Pennsylvania in 1935, some modifications had been made, including a built-up coal bunker. W. R. Osborne photograph.*

area, a short turntable was installed at the end of the platform.

Far less publicized but part of the same program, B & O simultaneously started a connecting bus service from Newark, New Jersey, meeting its trains at the Jersey Central's Elizabeth station. Newark was the state's largest city and on the Pennsy main line; by rail, B & O could serve it only indirectly through CNJ branch line trains from Elizabethport and Jersey City. The Newark buses were the same as the New York fleet, but were operated by New Jersey's Public Service Transportation (after 1928 called Public Service Coordinated Transport). In addition to the bus connections, B & O operated a direct Newark-Wash-

Photos this page: *Trainside to shipside to curbside: B & O train connection "motor coaches" met all trains at the Jersey City terminal platform, rode the ferry to New York (obviously a composite photograph, but not far from the real thing), and unloaded at terminals such as this at 191 Joralemon Street in Brooklyn. B & O Museum Collection, B & O photographs.*

Between Train-Side at Jersey City and the Coach stations at Waldorf-Astoria, Pershing Square and Brooklyn.

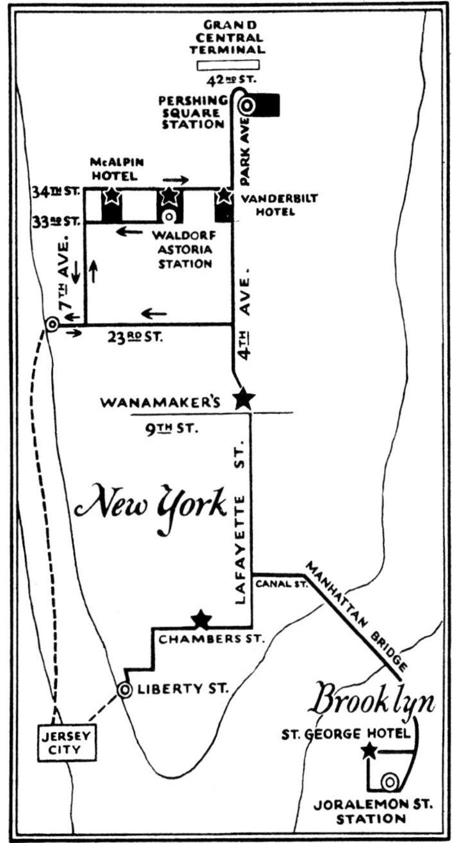

Motor Coach Stops — *New York*

23rd Street Route—McAlpin Hotel, Broadway and 33rd-34th Sts.; Waldorf-Astoria Hotel Station, 33rd-34th Sts. just west of Fifth Ave.; Pershing Square Station, 42nd St. near Park Ave. —directly opposite Grand Central Terminal.

Liberty Street Route—Consolidated Ticket Office, 57 Chambers St.; Wanamaker's, 4th Ave. and 9th St.; Vanderbilt Hotel, 4th Ave. 34th St.; Pershing Square Station, 42nd St. near Park Ave.

Brooklyn—Liberty Street Route—Motor Coach Station, Central Building, 191 Joralemon St. near Court St.; St. George Hotel, 51 Clark St. corner Henry St.

The three original motor coach routes. In 1929 a fourth service was added to Columbus Circle at 59th Street. J. J. Snyder Collection.

ington sleeper, which was switched in and out of the overnight train at Jersey City.

Changeover day came August 29, 1926 when B & O P-5 Pacific No. 5204 arrived in Jersey City with train 528, the overnight run from Washington. Two days before the trip Willard had run a press special, and afterwards B & O passenger traffic people were brought up for a firsthand look. It was the beginning of a concentrated public relations and advertising effort that would last thirty years or so.

As the new P-7 Pacifics began arriving, train schedules were expanded and speeded. B & O returned to its old pattern of eight through round trips. Some of these carried through coaches for Chicago and Cincinnati (and later St. Louis) in addition to the traditional through sleepers. By May 1927 most trains were covering the rail portion of the Washington run in five hours with one, the *National Limited* (the old *Royal Limited's* successor) doing so in a record four hours 45 minutes. All of that, of course, was now measured from Jersey City; including the bus trip from the 42nd Street terminal, the fastest time was five hours forty minutes. By then several Pennsy trains were scheduled for five hours flat or five hours ten minutes between Penn Station and Union Station in Washington.

By the fall of 1929 service had grown to ten trains each way, an all-time service peak, and included a new direct competitor for the *Congressional Limited*, the *Columbian*. Inaugurated May 29, 1929 as B & O's fastest schedule yet, the *Columbian* left both New York and Washington in mid-afternoon and ran terminal-to-terminal in four and one half hours. Passengers were landed at 42nd Street in an even five hours; quite respectable, although the *Congressional's* time to Penn Station was four hours 25 minutes, including the Manhattan Transfer engine change. A direct descendant of the *Royal Limited*, the *Columbian* was B & O's showpiece train and carried coaches, a club-lounge, diner, parlor, and parlor-observation cars.

The motor coach services also expanded to cover more of New York City and establish a stronger B & O presence. A third bus route was added November 24, 1926, using the Liberty Street ferry and crossing the Manhattan Bridge into Brooklyn. Brooklyn motor coaches stopped at the Hotel St. George and a B & O station-ticket office at 191 Joralemon Street opposite the Borough Hall. And on December 17, 1928 the 42nd Street station was moved to a new showcase terminal in the newly-built 54-story Art Deco-style Chanin Building, also opposite Grand Central at Lexington Avenue. Always B & O's busiest New York station, the new 42nd Street quarters were built around an elaborately decorated indoor loading/unloading facility, complete with bus turntable, and included a waiting room/ticket office fitted with marble wainscoting, Art Deco lighting fixtures done in bronze and chromium, leather sofas, and writing desks. Finally, on August 26, 1929 a route to Columbus Circle was started, using the 23rd Street ferry and 8th Avenue. From then on B & O operated a total of thirty buses on four separate New York routes, advertising fourteen stops — plus the two-bus Newark service. Whether through stubbornness or optimism, B & O was now well dug in.

As always, B & O passenger salesmen found virtue in adversity. Passengers from New York were regularly reassured in timetables and advertising: "When you step aboard the coach you've *made your train!*"; "No traffic or baggage worries," and "Your ticket includes this service.... NO ADDITIONAL CHARGE." Somewhat more dubiously, early 1930s timetables carried the slogan "The Open Air Route" — presumably favorably comparing B & O's

Photos this page: *The ultimate B & O motor coach terminal was opened in the Chanin Building at 42nd Street in 1928. The palatially-decorated inside loading area included a turntable (foreground). The bank-like lobby opened directly to 42nd Street and to connecting under-street passageways to Grand Central Terminal, the Commodore Hotel, and the IRT subway. B & O Museum Collection, B & O photographs.*

Photos this page: Wilsmere Yard, outside Wilmington, was rebuilt in 1918 to take pressure off Philadelphia's congested and awkwardly-located facilities. B & O Historical Society Collection, B & O photographs.

ferry ride across the Hudson with the Pennsy's tunnels to Penn Station.

While the New York passenger service probably got an undue share of management attention, World War I and the 1920s were active times for Philadelphia and New York freight traffic too. Wartime loads quickly accentuated the operating shortcomings of the East Side yard in Philadelphia. Some relief came in 1918 when the Grays Ferry Avenue tunnel was enlarged to accommodate four main line tracks in the congested area where freights had to stop and reverse direction. And to avoid East Side wherever it could, B & O also rebuilt its Wilsmere yard outside Wilmington in 1918. Wilsmere was extensively used to switch the through New York freights; during the World War I era, in fact, B & O and Reading freight crews were pooled to handle trains east of there, and many freights ran through Philadelphia without stopping.

Also booming was the Bethlehem Steel mill and shipyard at Sparrows Point outside Baltimore. Since the complex was first built in 1889, B & O had reached it through the short branch to Colgate Creek, where both it and the PRR interchanged with the steel company's own railroad, the Baltimore & Sparrows Point. By 1918 this layout was overwhelmed and the entire pattern of access to Sparrows Point was revised. B & O built a new five-mile line, originally called its Patapsco Neck branch, running directly into Sparrows Point by way of Highlandtown and Dundalk. Along the branch was built another large new traffic source, the Army's Camp Holabird depot. At the same time the Pennsy took over the old Baltimore & Sparrows Point line as its own entry to the Bethlehem complex.

At the war's end B & O optimistically planned a major rebuilding of its Philadelphia Division to smooth out some of the many curves and up-and-down grades. A 1920 plan

Darby, Pennsylvania was once a major suburban stop 6.5 miles from Philadelphia, and also was reached by the Philadelphia Rapid Transit Company's Darby streetcar line. The B & O-PRT crossing looked like this in 1916, with the railroad station at the left and a PRT trolley heading toward Philadelphia. The station is long gone, but the crossing remains today — now the only active railroad-streetcar crossing left in the United States. B & O Museum Collection, B & O photograph.

proposed a series of minor grade revisions from Bay View (Baltimore) to Havre de Grace, ten miles of completely new line east of the Susquehanna to bypass Foy's Hill, and several shorter realignments at Newark, Wilmington, and Carrcroft, Delaware. A final major element in this plan was to build a new freight bypass between Eddystone, Pennsylvania (east of Chester) and East Side Yard in Philadelphia. This line would run through the gradeless lowlands east of both the B & O and PRR main lines, roughly following Darby Creek and the Reading's Chester branch; it would cross the Schuylkill into Philadelphia at the south end of East Side yard, providing a straight-on entrance to the yard.

At the same time, elaborate USRA-inspired plans to alleviate the perennially-congested B & O and PRR routes through Baltimore were debated. Various methods of coordinating the two main lines were suggested as ways of speeding traffic flow through the long double-track tunnels, largely by creating all-passenger and all-freight routes. Most ambitious was a new line to be called the Patapsco & Susquehanna, an "outer belt" freight route which would branch off the Pennsy and B & O lines south of Baltimore, swing far around to the west and north sides of the city, and tie into the B & O's Philadelphia line at Van Bibber, Maryland. Eastward from Van Bibber, Pennsy freights would use the B & O as far as Havre de Grace, where a flyover junction was planned to rejoin the PRR main line. The Patapsco & Susquehanna's exact route was never precisely set, and several variations were proposed, including a separate line all the way to Havre de Grace. In any event, postwar problems deferred action on any of these ideas, including the B & O's own rebuilding plans, and ultimately few ever materialized.

One which did was a drawn-out program to coordinate, relocate, and elevate both the B & O and Pennsylvania freight lines in South Philadelphia. The two railroads had active, grade-level branches across the "neck" on the far south side of the city to reach their piers and freight facilities along the Delaware River. The B & O's line occupied the center of Oregon Avenue for much of its length and the Pennsy had a surface line located between Packer Avenue and Pollock Street. Both routes dated to the 19th Century (the Pennsy's to 1861) when much of the area was a swampy no-man's land. But after the turn of the century, the city began spreading south and the railroad tracks had become a barrier to development. By 1913 political pressures had built to the point where the city government and the two railroads started serious planning to relocate the "neck" lines to a grade-separated route on the extreme southern fringe of the area. The B & O and Pennsy agreed to share the route and the city agreed to share the cost.

Between the problems of creating joint legal agreements, designing and coordinating the three-way financing, and the delay of World War I, the project did not start until the early 1920s; it then dragged on through the entire decade. When finally finished in 1931 the joint route (with a separate track for each railroad) was located mostly on a fill looping around the city's south end and bordering the Navy Yard.

The B & O and Reading also jointly built a large perishable products terminal near the Delaware River waterfront at Jackson Street between Delaware and Weccacoe Avenues. Philadelphia traditionally was a major regional food distribution center, and the new facility was designed to receive fresh fruits and vegetables, store them if necessary, and provide quarters for daily auctions. When completed in 1927 the terminal included team unloading tracks for 360 cars, auction and private sales buildings, and a large cold storage warehouse. The terminal did more than merely create a much-needed freight revenue source in Philadelphia; it made the B & O a significant supplier for the city's food businesses, moving in trainloads of both southern and western perishables. (The terminal's joint owner, the Reading, had a smaller perishable business, and

The new Philadelphia Perishable Products Terminal rises along Delaware Avenue in 1926. In the foreground are team tracks for unloading fruits and vegetables directly into trucks. B & O Historical Society Collection, B & O photograph.

B & O sometimes was accused of using its Reading stock ownership to force that railroad to share the cost.)

B & O also operated a modest tug and carfloat "navy" in Philadelphia to handle some of the same types of work as in New York. Several New York-style pier stations were maintained along the Delaware River, served by carfloats from a float bridge at Dickinson Street; the floats also were used to interchange freight with the Reading's Atlantic City Railroad lines at Camden, New Jersey.

Once in business, the terminal was a constantly active place, handling auction sales, direct sales, and storage. Here some shipments are opened for pre-sale inspection. B & O Historical Society Collection, B & O photograph.

Pioneering diesel No. 1 had been renumbered to 195 when this shot was taken in 1955, but it was still working the West 26th Street terminal and float bridge. H. H. Harwood, Jr. photograph.

In addition, B & O was permitted to use a section of the Philadelphia Belt Line track on Delaware Avenue along the waterfront as far north as Callowhill Street, an arrangement dating to 1911. This picturesquely tangled street trackage was interwoven among Pennsy lines and served industries, warehouses, and piers along the river; both the B & O and PRR assigned small 0-4-0 switchers to the area to negotiate some of the tight siding curves. The area, in fact, was briefly the home of B & O's only gasoline-powered switcher, a tiny eighteen-ton Plymouth product bought in 1926. (Some 0-4-0s also were used to switch Wilmington industries.)

Meanwhile in New York, a much more portentous new form of motive power showed up at B & O's little 26th Street yard, a diesel-electric switcher. Again the B & O became a pioneer, and again not by choice. After twenty years of trying to eliminate railroad steam locomotives in New York City, the state passed the Kaufman Electrification Act in 1923, requiring electric power within the city limits. A 1926 deadline was set, which was later extended to 1931; the law also was modified to allow other types of non-steam power, such as internal combustion engines. At about the same time, the partnership of General Electric and the Ingersoll Rand Company, a diesel engine builder, had developed a practical, reliable diesel-electric switcher which would meet the law's requirements. For small-scale operations such as the B & O's, the diesel obviously was preferable to the expense of electrification — although it was yet unproven. Along with several other New York-area railroads, B & O tested the GE-IR demonstrator in late 1924.

The outcome was an order for a single sixty-ton 300 hp boxcab diesel. Delivered in late 1925, the B & O unit (again numbered "1") was the second of its type off the production line, preceded only by the Jersey Central's pioneering No. 1000. Boxy, bland-looking, and determinedly utilitarian, the CNJ and B & O units were the first diesels put in regular railroad service, and were the forerunners of the diesel revolution. The true revolution, however, did not start until more than ten years afterward; in the meantime most railroads and their suppliers viewed the scattering of small diesel switchers in New York and elsewhere as merely "special situations." Nonetheless, B & O No. 1 (later renumbered 195 and then 8000) chugged reliably between the 26th Street yard and the carfloat slip for 33 years. (Currently it is preserved at the National Museum of Transport at St. Louis, while CNJ No. 1000 resides at the B & O Museum in Baltimore.)

Unfortunately the 26th Street yard was not B & O's only operation affected by the Kaufman Act. Although still semi-rural and isolated, Staten Island was a part of New York City and the law applied there too — especially to passenger trains. (Freight service on Staten Island apparently was treated more casually; steam continued to work the yards and transfer runs until 1946.) At this time the SIRT's dense passenger operations still consisted of the light 4-4-0 camelbacks and dinkier 2-4-4Ts hauling their wooden coaches. By 1921 the compact SIRT system was carrying more than thirteen million people a year, but at best it was a questionable moneymaker. Nevertheless B & O raised no objections to its new legal obligation and immediately started work on a full-scale rapid transit-style electrification estimated to cost $5 million.

Although it was a large capital outlay for a dubious business, B & O was willing because it had a larger vision in mind. Since 1919 the city, the B & O, and several other New York-area railroads were involved in an ambitious plan to connect Staten Island with Brooklyn by rail. As the

SIRT's No. 27, a light 4-4-0, heads out of St. George terminal with a Tottenville train in 1919. Smithsonian Institution Collection, C. B. Chaney photograph.

project slowly evolved, the city government intended to sponsor a form of union belt railroad which would connect the railroads on the New Jersey side of the harbor with freight terminals and piers in Brooklyn and Queens, including a planned new port development on Jamaica Bay. Other parts of the plan included a passenger and freight connection to the Hell Gate Bridge route to the Bronx and New England, as well as a link to the city subway system in Brooklyn. The key to the project would be a pair of city-financed tunnels under The Narrows between St. George in Staten Island and the Bay Ridge area of Brooklyn. One double-track tube would carry freight trains, the other

After 1925 the St. George rail-ferry terminal looked like this. Al Gilcher photograph.

passenger. B & O obviously would be the major beneficiary of the planned freight links, since it would have direct rail access to the existing and future terminals on the Long Island side of the harbor. In addition its SIRT subsidiary would be tied into the New York subway system, joining the Brooklyn-Manhattan Transit Company's Fourth Avenue line at its 59th Street station in the Bay Ridge section.

By 1924 the city had begun active construction work on the Narrows tunnels, although nothing yet had been done on any connecting railroad lines. Based on what seemed to be an assured connection with the BMT subway, B & O designed its Staten Island electrification and cars to be compatible with the BMT's. This included a 600-volt DC third-rail distribution system and 67-foot-long steel multiple-unit cars equipped with three power-operated sliding doors on each side. But to replace the SIRT's old semaphore and Hall-patent "banjo" signals, B & O decided to install its own newly-developed color position light design — the first permanent installation of what gradually would become B & O's standard signaling system.

The Staten Island electrification project went quickly and smoothly. Work started in August 1924 and the first segment, the St. George-South Beach branch, opened in June 1925. The St. George-Tottenville main line followed July 1 and the Arlington branch along the north shore was finished in late December. A total of ninety electric motor cars and ten trailers (later also converted to motors) took over all passenger services; although later severely diminished in number, they remained the SIRT's backbone for 48 years.

But the Narrows tunnel never got beyond the digging of shafts at its two ends. Political pressures (reputedly inspired by the Pennsylvania Railroad) led to the elimination of the freight tunnel from the plan in 1925. Beset by other political problems and its awesome cost, the entire project had died by 1926, leaving the B & O without its hoped-for freight and subway connections, but with its very own peculiar variety of urban rapid transit system. The Staten Island's subway-style trains ran through woods and meadows, stopping at small wooden way stations with names like New Dorp and Great Kills; they also threaded their way through freight yards and past piers, factories, and warehouses. As the years went on the odd railway lost much of its bucolic flavor, particularly after an end-to-end grade crossing elimination project was begun in the 1930s.

New York's "rural rapid transit" still looked the part in 1954 when this Tottenville-St. George train paused at New Dorp station. H. H. Harwood, Jr. photograph.

By 1927 eight new steeple-cab electrics had replaced all of the Baltimore Belt Line's early power. Three of them grind uphill north of Falls Road with a freight in 1943. C. T. Mahan photograph.

Yet even today, under state ownership, it remains partly a Toonerville rapid transit.

Beside its through New York passenger trains and its incongruous Staten Island operation, B & O also rather listlessly plugged away at local passenger services in the area between Baltimore and Philadelphia. Always a sparse market, it had become progressively more so through the 1920s as automobiles, buses, and local trolley lines drained away business. The 34 local and commuter trains of 1888 had dropped to twenty by 1916. By 1927 there were only twelve, six each way, most of them scheduled for commuters into Philadelphia or Baltimore. One of these six round trips doubled as a Philadelphia-Wilmington local and as the sole remaining passenger service over the scenic Landenberg branch between Wilmington and Landenberg. Not shown in most late-1920s timetables was also a special commuter service for employees of the new Baldwin Locomotive Works complex at Eddystone, Pennsylvania. Baldwin gradually had been moving its plant from Broad Street in downtown Philadelphia to Eddystone during the 1920s and completed the transition in 1928. B & O accommodated Baldwin's Philadelphia workers with trains originating at 31st and Girard in Philadelphia, a Reading local station just west of Park Junction, and running directly into the sprawling Eddystone plant twelve and one half miles south. To get into Baldwin they used a short section of the old Crum Creek branch which directly entered the plant.

In Philadelphia B & O made a few tentative moves to replace its original 24th and Chestnut Streets station. The Furness building, exciting when it was finished in 1887, soon became unfashionable and "ugly." Several designs were developed for a modern combination station-office building in Art Deco style, but none got farther than the sketch stage. Another plan, considered by the B & O in 1925 but quickly rejected, would have used the former Baldwin Locomotive Works plant property at Broad Street near Spring Garden as a joint B & O-Reading station site.

By the close of the 1920s, the one-time Royal Blue Line, now with no distinctive name or symbol of its own, was a mixed picture of prosperity, uncertainty, and decay. Freights, now headed by Q-class 2-8-2s, hauled heavy loads of perishables and coal to New York and Philadelphia, returning with merchandise and less-than-carload shipments. The New York train/bus service could hardly compete head-on with the Pennsylvania, but it developed its own market niches. One was traffic to and from points on B & O's main line west of Washington, particularly in western Pennsylvania and eastern Ohio; another was suburban New Jersey business served through the CNJ stations at Plainfield and Elizabeth. The New York bus operation also gave the railroad an effective weapon for soliciting tour groups. Specially-assigned B & O motor coaches could pick up and deliver tour parties directly at their hotels anywhere in midtown Manhattan, in effect offering a private taxi service between hotel and train. Special groups grew to become a major part of B & O's New York passenger business.

But the old weaknesses persisted. B & O's back-door territory between Baltimore and Philadelphia developed slowly or not at all, and motor competition (both for passenger and freight business) took away much that had been there. Even by 1929, many of the pretty way station buildings no longer had agents. The loss was not all bad; B & O was spared the burden of heavy commuter traffic which later plagued the Pennsy. But unfortunately it was also spared much industrial development. New York and Philadelphia were healthy but constricted freight markets. In New York B & O was limited to whatever it could reach by water within the harbor lighterage limits, plus any on-track industry which it could develop on Staten Island. At its

The Columbian soon after its 1929 introduction. B & O Museum Collection, B & O photograph.

peak in Philadelphia, B & O reached only three percent of the city's railroad industrial sidings, and depended heavily on its perishables business and industries in the south Philadelphia peninsula.

All this time the Pennsylvania continued to move and build. Back in 1915 it had completed its first large suburban electrification project, thirty miles between Philadelphia and Paoli, using a newly-adopted 11,000-volt AC catenary system. In 1928 its aggressive new president, W. W. Atterbury, announced that electric wires would be strung between New York and Wilmington, and by 1930 work was under way or completed on several sections of that route. Not much clairvoyance was needed to predict that the entire New York-Washington line would soon be electrified. B & O studied its own electrification between Baltimore and Philadelphia in 1928 and again in 1932 but, predictably, nothing came of it.

Then came the disaster of the Depression which, paradoxically, produced a second coming of Royal Blue Line luxury, innovation, and image.

Royal Blue Reborn

The 1930s

The Depression hit B & O with a special fury. By 1933 the company's total revenues had plunged 46 percent from their 1929 level and were only half of the 1926 peak. Large but ill-timed investments in other railroads — notably the Western Maryland, Chicago & Alton, Buffalo, Rochester & Pittsburgh, and Buffalo & Susquehanna — further drained its treasury. Beginning in 1932 Willard was forced to negotiate some government-backed loans from the Reconstruction Finance Corporation and Public Works Administration to stave off bankruptcy.

In the meantime the railroad pruned wherever it could, including the newly-expanded New York-Washington services. By April 1932 the ten through-New York round trips had been scaled back to the traditional eight, although the recently-established *Columbian* survived as the flagship train and was speeded up to four hours 22 minutes by early 1935. One of the two Philadelphia-Washington expresses was cut, leaving only the morning "businessman's commuter" schedule from Wayne Junction, which returned from Washington in late afternoon. On September 28, 1930 the Landenberg branch lost its sole passenger train, and in December 1931 all remaining local passenger services on the Baltimore-Philadelphia line were reduced to two commuter runs: one between Philadelphia and Aiken, Maryland and the other between Aiken and Baltimore. The terminal of these two trains later was moved to Singerly, Maryland, thirteen and one half miles

In a typical early 1930s scene, a New York-Washington train loads at Wilmington's Delaware Avenue station. By this time, "President" Pacifics like the 5317 headed virtually all through trains. B & O Museum Collection, B & O photograph.

Lesser trains got lesser power. In 1941 the traditional morning Philadelphia-Washington semi-express rolls through Rosedale, Maryland behind Pacific 5074, a 1911 P-1a. M. A. Davis photograph.

east of Aiken. Gas-electric "motor" cars and trailers handled these two skeleton schedules for most of their remaining lives.

As the Depression deepened, additional moves were made to cut costs. In an attempt to rationalize the New York-Philadelphia services being operated by both B & O and the Reading-Jersey Central, some Reading-CNJ runs were cut and B & O trains were used to help fill in at the intermediate points. Beginning in November 1933, five B & O trains each way were routed over the original line through Jenkintown, Pennsylvania to accommodate Reading passengers between the Philadelphia suburb and New York. Some B & O trains also stopped at West Trenton, another Reading local point. The Jenkintown detour added time to the B & O runs, and it was ended in April 1935 when conditions began to improve slightly. A permanent Depression casualty, however, was the Washington-Newark sleeper run, which was discontinued in April 1935.

Aside from these moves, however, Willard not only stayed in the expensive and heavily one-sided New York competition but fought even harder. Hope, tradition, and corporate pride probably were the compelling reasons. But there were more practical ones too: New York was the corporate headquarters for many of B & O's large freight customers, and the home of the financial institutions which supported it. The ability to provide a first-class passenger service at their doorsteps was considered a critical selling point.

Faced with the certainty of the Pennsylvania's electrification of its Washington services (Atterbury finally announced the project in February 1931) and his own severely limited resources, Willard had to devise a strategy which somehow would challenge the speed and cleanliness of Pennsy's electrification — but at a much cheaper price. His answer was to combine B & O's "personal service" tradition with the newest and most significant technical innovations being developed by the railroad industry. B & O would be the first to adopt virtually every new innovation which promised greater speed, passenger comfort, and convenience; and in the 1930s there were many. The decade thus was one of the most exciting in B & O's history, indeed the entire industry's history, and much of that history would be made on the New York line.

About fifteen minutes out of Jersey City, the St. Louis-bound Diplomat rumbles through the Jersey Central's old station at Elizabethport, New Jersey in 1937. A key junction point for the CNJ, Elizabethport was disdained by B & O trains. Smithsonian Institution Collection, C. B. Chaney photograph.

First was air conditioning. The Pullman Company had made two serious but unsuccessful attempts to air condition sleepers in 1927 and 1929. B & O picked up on the idea and in the summer of 1929 had Willis Carrier install a test air-conditioning unit in one of its coaches. Carrier's equipment worked well enough to encourage Willard to keep moving. In April 1930 Carrier equipment was installed and tested in a new "Colonial"-series diner, the *Martha Washington.* This time it was deemed suitable for regular service, and in May the *Martha Washington* was added to the *Columbian,* the country's first air-conditioned car in regularly scheduled operation. Following further tinkering with the machinery, B & O achieved a fully reliable system and began equipping a complete train of air-conditioned cars for its *Columbian.* On May 24, 1931 the refitted *Columbian* made its first trip, and was immediately billed as the world's first fully air-conditioned train. It was an instant success. The Pennsylvania's mind was on electrification, and its own air-conditioning program lagged behind by more than a year; it was not until 1933 that Pennsy air-conditioned its New York-Washington trains. In the meantime, the *Columbian's* business swelled and extra parlor cars had to be added.

Reclining seat coaches also were introduced in 1931, and in 1932 the *National Limited* began carrying through air-conditioned sleepers. By early 1933 three of B & O's eight New York trains were fully air conditioned: the *Columbian, Capitol Limited,* and the *National Limited.* Most others carried air-conditioned diners or lounge cars, and all advertised "individual seat coaches." The air-conditioning

Fully air-conditioned, the premier Columbian speeds east on the Reading at Roelofs, Pennsylvania (near Yardley) in 1936. W. R. Osborne photograph.

Surely one of the daintiest 4-6-4s ever built, the No. 2 Lord Baltimore was custom-designed for its lightweight train. Shown in shining new condition at Jersey City in 1935, it was indeed fast but unadaptable to other services. A. P. Formanek photograph.

program on the New York-Washington line was completed in June 1934.

More creativity was needed, though. The Pennsylvania's Washington electrification was scheduled for completion in 1935 and Willard had to be ready for that. So while the Pennsy was stringing its wires, Willard plunged into a project which would pull together all of the latest concepts of passenger train and motive power design, many of which were scarcely proven. In mid-1934 he negotiated a $900,000 Public Works Administration loan which would be used to make B & O's New York-Washington line an industry-wide proving ground for various types of lightweight train construction and high-speed steam and diesel power. The Union Pacific's and Burlington's 1934 lightweight motor trains had dramatically demonstrated both the passenger appeal of streamliners and the practicality of internal combustion power for high-speed passenger service. But both the UP's "M10000" (later *City of Salina*) and the Burlington *Zephyr* were merely three-car articulated train sets of limited capacity and power. The next step was to extend the ideas to more "normal" railroad operations, meaning full-length lightweight streamlined trains and higher horsepower road diesels. And, remembering that in 1934 the diesel was only a promising but still-questionable newcomer, there was also an opportunity to demonstrate steam power's potential on the same type of trains.

Thus the B & O ordered two full streamlined train sets to be built of different types of lightweight materials, plus one diesel locomotive; in addition it ordered its own me-

In the seemingly eternal tradition, "Miss Royal Blue" blesses the new train on its maiden trip from Washington on June 24, 1935. Actually, she was Marie McIntyre, daughter of President Franklin Roosevelt's secretary, Marvin McIntyre. E. H. Hinrichs Collection.

Photos this page: *Inside and out, the lightweight Royal Blue was sleek and modern, but also rather austere.* ***Left:*** *Inside its parlor-lounge-observation car, somewhat fusty, homey furniture sat in coldly functional surroundings. E. H. Hinrichs Collection.*

chanical forces to design and build steam locomotives which could match the diesel's performance.

The car order went to American Car & Foundry and totaled sixteen cars. One eight-car set was to be built of Cor-Ten steel, a newly-developed alloy, the other was to be of aluminum on a lightweight steel framework. As originally planned, both trains consisted of six cars (with two spares for each), and both were to work the New York-Washington route, allowing side-by-side comparisons.

For motive power, B & O provided a specially-designed light high-speed 4-4-4 steamer built at its Mt. Clare shop in Baltimore. General Motors' Electro-Motive Corpo-

The solid blue streamliner heads east at Ivy City in Washington, DC in September 1936. Behind it are the newly-strung electric wires of the paralleling Pennsy, one of the major motivations for building the streamliner. W. R. Osborne photograph.

Photos this page: *The Royal Blue at its two terminals.* ***Above:*** *Carefully watched over in Washington. Bruce D. Fales photograph.* ***Below:*** *Alongside its bus connection at Jersey City. Ted Gay photograph.*

ration subsidiary was to build an 1800 hp road passenger diesel, identical to two prototypes which it planned to demonstrate to the railroads. Perhaps already recognizing the limitations of its 4-4-4 steam prototype, B & O added another locomotive as the project progressed, a high-speed 4-6-4 of similar design. The two steamers were not streamlined, but were designed with a clean, lean English appearance reportedly inspired by the visit of the Great Western Railway's *King George V* at B & O's 1927 Centennial celebration. The 4-4-4 originally was to have an inverted bathtub-style streamlined shroud; this had been partially installed when Willard saw it and emphatically ordered it removed. Both were high-pressure locomotives designed to operate at 350 psi and were fitted with water tube fireboxes. They rode on 84-inch drivers, the largest ever used on a B & O locomotive. They were intentionally light, however, and

specifically designed for speed and economy with lightweight trains. The 4-6-4, for example, weighed thirteen percent less than the P-7 Pacifics used on the New York runs, and even with its tender booster working, it produced eighteen percent less tractive effort.

Plans changed before the trains were delivered. Instead of operating both on the New York-Washington line, it was decided to assign the Cor-Ten steel train to Chicago-St. Louis service on B & O's subsidiary, the Alton Railroad, where it was named the *Abraham Lincoln.* Along with it went the 4-4-4, now called the *Lady Baltimore.* The aluminum train was assigned to the New York-Washington run as planned, hauled by the 4-6-4 *Lord Baltimore.* The single diesel alternated between the New York and the St. Louis trains.

Reaching back to its gilded age, B & O named its new New York-Washington streamliner the *Royal Blue.* Actually it was the first time the name was specifically applied to a single train; in the past, "Royal Blue" had been applied generically to the entire fleet of trains. The cars and locomotives were painted a deep solid blue, comparatively colorful compared to the conventional olive green, but certainly not as eye-catching as the yellow and tan of Union Pacific's new streamliners, the Burlington *Zephyr's* shining stainless steel, or the orange and maroon of the Milwaukee Road's new *Hiawatha.* Both the *Royal Blue's* name and its color scheme were somehow symbolic of Daniel Willard's later years, a curious mixture of radical innovation and deep conservatism.

Oddly, the most radical component of B & O's streamliner program, the diesel locomotive, looked the most retrogressive, especially next to its Union Pacific and Burlington forebears. A plain, utilitarian boxcab design, the high-speed diesel resembled nothing more than a stretched yard switcher; its squared-off lines looked absurdly incongruous at the head of the sleek blue streamliner. Strangely enough, the normally savvy General Motors, who was responsible for the styling, was surprisingly slow to adopt streamlining for its locomotives.

Certainly more appropriate for the train was B & O's own 4-6-4 No. 2, the *Lord Baltimore.* Although not fully streamlined either, its smooth English lines with long flush pilot made it look fleet — which it was — and its enclosed cab, streamlined tender, and full-width rear tender diaphragm blended with its train.

The *Lady* and *Lord Baltimore* were successively tested February 26th and 27th, 1935 to see how they might perform on the *Royal Blue's* demanding schedule. Both took turns on a four-car consist of conventional steel (but comparatively lightweight) equipment, and both easily bettered the planned running time between Washington and Philadelphia. In rainy winter weather the dainty *Lady* slipped several times; with its speed intentionally held to a 75 mph maximum, it nonetheless hit 81 at one point. With no speed restriction and a clear, cold day, the *Lord* got up to 90 mph eastbound and 88 mph on the return trip.

The new *Royal Blue* made its maiden revenue trip June 24, 1935 hauled by the *Lord Baltimore*; the diesel, which B & O numbered "50", was not delivered until August. Immediately advertised as "The World's Most Modern Train" (a bit of a half-truth), the *Royal Blue* was the fastest schedule yet attempted on the route — four hours between Washington and Jersey City, 22 minutes faster than the premier *Columbian.* (For several years after, the train did not make an eastbound passenger stop at Baltimore's Camden station.) Between Washington and 42nd Street in New York, using the bus connection, the total time was only four and one half hours. As it turned out, the *Royal Blue's* four-hour schedule was the fastest ever managed by the B & O-Reading-Jersey Central route, and was maintained until the service ended in 1958. A three hour 45 minute schedule was proposed and almost operated, but Daniel Willard was reluctant to raise the train's speed limit.

The Victorian clock tower of the Jersey Central's 1894 Elizabeth station fades into the background as the Lord Baltimore eases the Royal Blue away from the station stop. H. W. Pontin photograph.

Photos this page: ***A stylist's nightmare, Electro-Motive-built diesel No. 50 nevertheless had solid innards and helped start the diesel revolution.*** ***Right:*** *On one of its first trips, No. 50 waits with the westbound Royal Blue in Philadelphia in August 1935. W. R. Osborne photograph.*

The *Royal Blue* moved fast, however. On its first trip, Everett Thompson from B & O's passenger department timed many 80 mph stretches between Washington and Philadelphia; on the Reading's racetrack between Philadelphia and Bound Brook, the *Lord Baltimore* wheeled the train past Belle Meade and Weston at 92 mph, and at 96 mph by Hamilton, New Jersey.

The *Royal Blue's* schedule called for a full round trip each day, leaving New York in mid-morning and returning from Washington in late afternoon. Initially it consisted of eight cars: a baggage-mail car, three coaches fitted with 64 reclining seats, a diner-lunch counter car, two parlor cars, and a round-end parlor-lounge-observation car. Within a month of the train's inauguration, one of the coaches was quickly refitted as a buffet-lounge car to provide bar-lounge facilities for coach passengers. The following year the baggage-mail car re-emerged from Mt. Clare Shop as a baggage-coach combine, and in 1937 the B & O shop pro-

At about the same time, it takes the new "Blue" through Bayonne, New Jersey on CNJ trackage. A. P. Formanek photograph.

duced an additional coach built of welded lightweight steel.

The *Royal Blue's* June 1935 introduction date was not accidental. On February 10th of that year the Pennsy began phasing in its New York-Washington electric services, and on April 8th it announced that all trains were now electric powered. Also in April the Pennsy began receiving the first of its magnificent GG-1 class electric locomotives, unquestionably one of the finest examples of railroad motive power ever built. In September 1935 the *Congressional's* time was cut to three hours 45 minutes, a schedule B & O would never be able to match. And by the end of the next year the "Congo" was down to three hours 35 minutes, a record for New York-Washington time that stood until 1967 when, as a prelude to the *Metroliner* program, it was temporarily accelerated to three hours twenty minutes. But if the B & O could not match the speed of the Pennsy's GG-1s, Willard had made certain his new streamliner equaled them in the public's eyes as a symbol of progress.

Fast, modern, and luxurious though it was, the *Royal Blue* was not an unqualified success, nor was its new motive power. The diesel locomotive functioned well enough and proved itself economically superior to its steam competition. But the single unit's 1800 hp was marginal for the speeds demanded, particularly on the Baltimore-Philadelphia section. Reported Everett Thompson on one of its first runs, "the 'motor' is wonderful on the starts but does not do the job on the hills that the steam engine does." As the railroad quickly discovered, diesel horsepower drops off sharply at higher speeds, requiring additional power to match steam performance at those speeds. Furthermore, No. 50's four-wheel trucks tended to "nose" at higher speeds, causing a rough and unsettling ride. The *Lord Baltimore* was better suited to the train's power and speed needs and, on the evidence of many photos taken during the period, seemed to haul the train much more frequently than the diesel. It was occasionally supplemented by a P-7 or a 1913 P-3. But the *Lord* was never duplicated. Custom-designed for light, fast trains, it had little future when

The sleek Lord Baltimore's "protect" engine was ordinary, elderly Pacific No. 5104, a 1913 P-3 repainted solid blue. It is shown westbound on the Baltimore Belt Line at North Avenue in September 1936. Within a minute or so, it will brake to a stop at Mt. Royal Station. Smithsonian Institution Collection, C. B. Chaney photograph

B & O abandoned lightweights and even less after the diesel's teething troubles had been worked out.

The aluminum train itself, which was ten percent lighter than its Cor-Ten steel sister and 46 percent lighter than a comparable conventional train, reportedly rode roughly on B & O's curving Philadelphia line. B & O historian and one-time public relations official Lawrence Sagle later wrote: "Because of the tight-lock couplers there was a jerky motion in the train when No. 2 was on the head end and the engineer had the Johnson Bar in the corner [i.e. making fast time], it was difficult to carry on a conversation in the head cars because of the vibration." Others such as Everett Thompson dispute this as exaggerated, however. But the *Royal Blue's* worst problem undoubtedly was Daniel Willard himself. Despite his well-publicized sponsorship of the project, Willard, who turned 75 in 1936, disliked and distrusted lightweight passenger equipment.

In 1936 the New York Central made railroad industry news by rebuilding a set of conventional steel suburban coaches into an attractive, fast, modern streamliner it christened the *Mercury*. Perhaps inspired by the Central, Willard soon began planning a replacement for his lightweight train, using rebuilt standard heavyweight equipment. Less than two years after its inauguration the original *Royal Blue* was yanked from the New York service, shopped, and sent west to join its sibling in Chicago-St. Louis service on the Alton. In a somewhat baffling shuffle of names, the former *Royal Blue* became the *Abraham Lincoln* and the original *Abraham Lincoln* was renamed the *Ann Rutledge*. Along with it went diesel No. 50 and the *Lord Baltimore*. Whatever the lightweight train's problems were on the New York run, they apparently vanished in the midwest; the *Abraham Lincoln* lived a long life, operating into the late 1960s. The diesel eventually was given six-wheel trucks and mated with other units, solving its major problems. It too survived, working passenger assignments on the Alton and, after another rebuilding, ending its working career in local freight service. It was finally retired in 1959 and currently is preserved at St. Louis's National Museum of Transport. The *Lord Baltimore* was less lucky; displaced by diesels on the lightweight streamliners, it returned to the B & O and ended its life as a limited-service orphan before being scrapped in 1949.

But if the lightweight *Royal Blue* disappeared, the name and service did not. Replacing it on April 25, 1937 was another eight-car streamliner ambiguously advertised as the "Improved *Royal Blue*," running on the same four-hour schedule. At first glance it was just as sleek and considerably more colorful, a pleasing lighter blue and gray scheme with gold striping, created by industrial designer Otto Kuhler. Kuhler's inspired new *Royal Blue* color scheme seemed a perfect fit for B & O's Willard-era corporate personality — bright, attractive colors conservatively

Otto Kuhler's streamlined 5304 at Washington in November 1937. L. W. Rice photograph.

applied. To some, it was the handsomest railroad livery of all time. Later it was used for other B & O streamliners and eventually became the railroad's standard passenger car scheme.

But beneath the "Improved *Royal Blue's*" streamlining and paint were standard heavyweight cars of the 1925-1927 period, extensively refurbished by B & O's Mt. Clare Shop. The "new" train consisted of a baggage-coach combine, an eighty-seat coach with old-style walkover seats, two 54-passenger coaches with roomy reclining seats and ladies' lounges, a coach-buffet car, a diner-lunch counter car, one parlor car, and a blunt-end buffet-lounge-observation car. Compared to its lightweight predecessor, the new train had greater coach capacity and less parlor space. Except for the combine and diner, all cars were rebuilt from coaches.

Heading the new *Royal Blue* was another example of something old made to look like something new. The train was powered by a conventional P-7 "President" Pacific, No. 5304, which had been given a bullet-shaped streamlined shrouding also designed by Otto Kuhler. Basically blue with gold striping and chrome trim, No. 5304 carried the *Royal Blue* name painted in gold on its running board. Although essentially unaltered underneath, it had been slightly souped up by increasing its cylinder diameter by half an inch and raising boiler pressure from 230 to 240 pounds. Although probably more symbolic of the *Royal Blue* than any other locomotive, the 5304 actually had a rather brief and erratic career on the train; often a regular P-7 showed up on the head end until diesels regularly took over in 1938. And while steam sometimes substituted for the diesels through the mid-1940s, the 5304 was shorn of its streamlined shroud in 1939 and took its turn along with all the other P-7s. In 1946 it got a second streamlining for the Baltimore-Cincinnati *Cincinnatian*.

The heavyweight streamliner was deemed successful and became the prototype for the modernization of B & O's premier long-distance trains of the late 1930s and early 1940s, the all-Pullman *Capitol Limited* (1938), the *National Limited* (1939-1940), and the Washington-Chicago version

Fresh out of Mt. Clare Shop, the "Improved Royal Blue" is accelerating out of Elizabeth, New Jersey on the morning Jersey City-Washington leg of its daily round trip. Smithsonian Institution Collection, C. B. Chaney photograph.

of the *Columbian* in 1941. While his competitors enthusiastically entered the lightweight train era, Willard remained resolutely wedded to the old heavies. In addition to his personal prejudices, he was motivated by economy and a sincere wish to keep B & O shop forces busy and employed — still another instance of B & O's tradition of converting adversity into a virtue. The *Royal Blue* and its ilk were thus a unique blend of superficial modernity but basic homey fustiness. While Mt. Clare Shop tried to give the *Royal Blue* a straight, streamlined look, with flush roofs, full-width diaphragms, and skirting, the cars retained their old narrow windows, exterior rivets, six-wheel trucks and generally bulky, clunky look. Inside, superficially modern decors coexisted with ceilings which betrayed the old clerestory roof lines. Somewhat ironically the Reading, by then a B & O vassal, introduced a Budd-built stainless steel lightweight five-car streamliner on its New York-Philadelphia run in December 1937. Afterwards the Jersey City terminal

And farther west on the Reading, No. 5304 is wide open with the heavyweight streamliner at Glen Moore, New Jersey. W. R. Osborne photograph.

B & O's Columbian vs. the Reading's Crusader, or, more to the point, Daniel Willard vs. the oncoming lightweight train era. The Columbian's equipment was the first heavyweight Royal Blue train set. B. G. Saylor Collection.

provided an enlightening side-by-side contrast between truly modern equipment and B & O's version.

Following the successful inauguration of the heavyweight *Royal Blue,* Mt. Clare Shop was put to work on an almost identical train with the idea of making the *Royal Blue* and *Columbian* into twin streamliners. The second train set, also eight cars, was completed in late 1937; on December 9, 1937 it replaced the original set as the *Royal Blue.* The slightly older train then was rechristened the *Columbian* and put on a four-hour schedule opposite the *Royal Blue,* leaving Washington in mid-morning and returning from New York in late afternoon. Between them the two B & O streamliners did their best to compete with the Pennsy's morning *Judiciary* and afternoon *Congressional.*

The latest (and, as it turned out, last) version of the *Royal Blue* looked virtually the same as its predecessor and had essentially the same consist and accommodations, including the ladies' lounges in the coaches. Dining space was expanded, however; one of the coaches incorporated a fourteen-seat lunch counter, and the diner was rebuilt as a full diner. Interior decor was slightly more stylish too, with curved motifs in the coach lounges, diner, and observation-lounge, and included a set of center-facing curved seats and tables in the diner and a semi-circular padded bar in the lounge-observation. Lounge chairs were movable and upholstered in red, brown, and green frieze. Rust and red frieze dominated the coach and parlor seat upholstery, and the car interiors were a bright but neutral ivory and gray.

While the *Royal Blue* and *Columbian* were generally kept intact as integrated train sets, other streamlined heavyweight cars slowly appeared in the New York service. In 1938 the all-Pullman Washington-Chicago *Capitol Limited* was "modernized" with the same streamlining technique and color scheme; after September 1940 through streamlined *Capitol* sleepers operated to New York. Streamlining of the *National Limited's* Washington-St. Louis consist was completed in 1940 too, and its remodeled through coaches and sleepers were also mixed with conventional cars on the New York-Washington leg of its run.

If Willard wanted to forget lightweights, the diesel was something else. Learning from its mistakes with B & O No. 50 and its own demonstrators, General Motors redesigned the basic machinery with six-wheel trucks and a streamlined body. By 1937 it had a production version of the passenger diesel ready and B & O was one of the earliest customers; the first units off the line, in fact, went to B & O in May 1937. Like No. 50, each unit had twin Winton diesel engines totaling 1800 hp, but they were paired in "A" (cab) and "B" ("booster") combinations to produce 3600 hp. The first of these twin-unit GM-Electro-Motive model EA-EB sets (both units were numbered 51) immediately was assigned to the *Royal Blue,* with the second set going to the Washington-Chicago *Capitol Limited.* Apparently soon afterwards the original set also was transferred to the *Capitol* and it was not until early 1938 that diesels were more or less permanently assigned to the *Royal Blue.* Nonetheless, the "Blue" did receive the honor of being pulled by the country's first streamlined passenger diesel. By August both the *Columbian* and the overnight train also were advertised as normally diesel-powered. Painted in a regal blue-gray-black-gold variant of Otto Kuhler's passenger car color scheme with some chrome trim, they blended well with their trains and, visually at least, clearly outclassed the Pennsylvania.

As the Royal Blue Line began operating some of the country's earliest streamlined diesels, it also was a proving

Photos this page: *"The most famous face in dieseldom," the streamlined cab unit, first appeared in 1937 on the Royal Blue. And to many people, the design and styling of B & O's EAs were never equaled on other railroads or in later models.* ***Right:*** *An EA/EB set prepared to take the Royal Blue out of Washington in November 1938. Carleton Parker photograph.*

Approaching Baltimore, another pair roars past Relay, Maryland with the eastbound "Blue" in 1940. E. L. Thompson photograph.

The westbound Royal Blue at Camden Station's lower level platform about 1938. R. L. Wilcox photograph.

Photos this page: *Transition at North Avenue. Taken a month apart, these two mid-1938 views from the North Avenue viaduct show the eastbound streamlined Columbian climbing across the Baltimore Belt Line after its stop at Mt. Royal.* ***Right:*** *Two OE-class electrics tow P-7 No. 5307 (which also appears to be working) across the PRR crossing in June. B & O Historical Society Collection, W. R. Hicks photograph.*

ground for what the coal-hauling B & O hoped might be a steam-powered competitor for the diesel. During 1937 and 1938 the *Royal Blue* and various other New York-Washington trains were occasionally pulled by the impressive experimental 4-4-4-4 duplex designed by Motive Power Superintendent George Emerson. A forerunner of the Pennsylvania's famous (and, to some, infamous) T-1 and Q-class duplexes, the B & O's *George H. Emerson* had opposed cylinders to reduce wheelbase length and, like most B & O experimentals, was equipped with Emerson's water

In July, shining new EA/EB No. 55 is clearly in charge and needs no assistance. Smithsonian Institution Collection, C. B. Chaney photograph .

The revolution that never happened: The George H. Emerson brings the eastbound Royal Blue into Baltimore in May 1938. One B & O engineer, attempting to make up time east of Baltimore, pushed the Emerson over 100 mph, earning a gentle reprimand but the heartfelt gratitude of its designer. Smithsonian Institution Collection, C. B. Chaney photograph.

tube firebox. Its maintenance problems coupled with the diesel's superior performance brought early death to the *Emerson*; B & O wisely dropped the idea and avoided the Pennsy's disastrous mistake. After being exhibited at the 1939 New York World's Fair, the *Emerson* went back to the Jersey City run, then was later reassigned to B & O's western main lines between Washington and Willard, Ohio. It finally was retired in 1943.

Streamlining also came to the New York bus service and, in fact, slightly preceded the heavyweight *Royal Blue* and the diesels. Late in 1936 B & O completely reequipped its now-obsolete motor fleet, replacing the 32 front-engine Yellow Coaches of 1926 and 1929 with 30 White model 684s equipped with underbody engines. Mechanically similar to the latest city bus designs, the B & O's new fleet was fitted with custom-built Kuhler-designed Bender bodies, and incorporated an early attempt at bus air conditioning. Also designed by Kuhler, the "air conditioning" system consisted of ice blocks in the bus's rear combined with fans. The new motor coaches were painted in a blue-black-gold version of Kuhler's B & O passenger diesel scheme, with three heavy chrome belt strips circling the lower body. But most striking was a giant gold-painted version of Kuhler's newly-designed B & O Capitol dome logo on their V-shaped front ends. Those who saw or rode the motor coaches did not readily forget them; for as long as they were in service, they were some of the most distinctive vehicles on New York's streets.

A fifth and final New York motor coach route and terminal was added shortly afterward. On May 3, 1937 B & O opened a station at the new Rockefeller Center at 49th Street and Rockefeller Plaza. Furnished in usual B & O style with sofas and armchairs, it was sleekly Art Deco modern to match the new buses. When opened, its interior wall was covered with an enormous photo-mural of the *Lord Baltimore* with the lightweight *Royal Blue* train posed on the Thomas viaduct — an overpowering but already obsolete bit of decor, since the train had been banished from the New York run only a few days before the Rockefeller Center terminal opened.

A new passenger revenue source also appeared in 1937. Delaware Park racetrack opened at Stanton, Delaware, roughly midway between Wilmington and Newark, Delaware. B & O built platforms by the track, which handled many special race trains over the years — trains which continued running even after regular passenger service ended.

Color, comfort, streamlining, and diesels made the reincarnated Royal Blue Line exciting during the dreary 1930s, but less glamorous events also were occurring — some of them positive, some not. Just as the Depression rolled in, B & O significantly increased its financial interest in the Reading. Between 1931 and 1932 it spent $71.2 million to boost its Reading stock ownership to 42.2 percent, giving it clear working control of both the Reading and Central Railroad of New Jersey. (The New York Central still held its old twenty percent interest.) To help cement the relationship, B & O's Eastern Region General Manager, Edward W. Scheer, became a Reading vice president in 1932 and moved up to the presidency three years

Photos this page: *Precursors of B & O's diesel streamliner livery, the 1936 White motor coaches were the aristocrats of the urban bus world and helped introduce "air conditioning." B & O Historical Society Collection, B & O photographs.*

later. Simultaneously, Scheer was named president of the Jersey Central. But as before, both the Reading and CNJ kept their separate corporate identities and remained separately operated. Although B & O still lacked an absolute majority control and had to be cautious of minority stockholders, its New York entrance was now tightly secured, and it was assured of reasonably subservient working partners.

Unfortunately the arrangement also carried the seeds of doom. In the years after, B & O never increased its Reading stock control or attempted a merger, so its influence was never absolutely complete or certain. One early warning of future problems came in October 1939 when the Jersey Central went into the third bankruptcy in its history.

An early product of the tighter B & O-Reading-CNJ relationship was a rationalization of the B & O and CNJ marine operations in New York Harbor. The low traffic of the Depression made the expensive lighterage and carfloating even more so, and the two railroads decided to consolidate their harbor services. Under an October 1934 agreement the two railroads pooled all their marine equipment (which remained separately owned) and based the entire operation at Jersey City under Jersey Central management. For the B & O, cost saving was only part of its motives. From the beginning, Staten Island had been a less-than-ideal expediency. The St. George terminal was too small and was constantly congested. Furthermore, its location was relatively remote; for most New York City traffic, B & O had a four-mile disadvantage compared to the other major lines. But although the pooling agreement essentially abandoned St. George as a water transfer terminal, its facilities were maintained and it remained in the tariffs. As traffic flows and volumes changed, St. George occasionally came back to life — particularly during World War II. For most of its life after 1934, however, St. George was merely a secondary terminal, and Jersey City remained

B & O's primary New York water transfer point and freight terminal.

Much of New York City's outbound freight business consisted of less-than-carload (l.c.l.) shipments, which either were consolidated into full carloads by independent freight forwarder companies or were handled directly as l.c.l. by the railroads themselves. By the late 1930s trucks had begun cutting heavily into this business; besides competing with one another, B & O and its peers now had to try to match truck service. B & O was under particular pressure because of its more roundabout route between New York and western points. As a hopeful experiment, in 1938

Photos this page: *Rockefeller Center station, cosily modern in the new B & O tradition. B & O Historical Society Collection, B & O photographs.*

it introduced an expedited New York-Pittsburgh train carrying exclusively less-than-carload shipments. Running on close to passenger train timings, trains 117-118 were among the fastest freights of the late 1930s and a now-forgotten predecessor of today's premium "piggyback" services. B & O equipped a small fleet of steel wagon-top boxcars for high-speed operation, gave the trains first-class timetable status, and often assigned passenger power, allowing 75 mph speeds between Philadelphia and Washington. In December 1938 B & O passenger official Everett Thompson clocked westbound No. 117 at 77 mph on the straight stretch through Aberdeen, Maryland, trailing behind a P-1a Pacific.

Trains 117-118 proved to be short-lived, but did add brief spice to an otherwise ordinary freight service. By the late 1930s, B & O's sturdy Q-4 class 2-8-2s powered most manifest freights over the Philadelphia line, often doubleheaded. Between Philadelphia and Jersey City, husky Reading and Jersey Central 2-8-2s with wide fireboxes hauled B & O freights in solid blocks, usually picking them up or dropping them intact on the main line at East Side. Some B & O freights continued eastbound as far as

Despite diesels and Kuhler's bullet-nosed 5304, the Royal Blue and Columbian sometimes would show up behind a conventional olive green P-7, as was the case in this 1938 scene as the morning Columbian left Washington for New York. H. W. Pontin photograph.

the Reading's Belmont yard in Philadelphia to drop cars for other Reading points. Although the Jersey City freights were Reading and CNJ trains east of Philadelphia, B & O set their schedules and advertised them as "B & O" trains. Most through freights were scheduled over the Philadelphia line at night to allow after business hours departures from New York and Philadelphia, and early morning arrivals for inbound shipments.

Daniel Willard at last retired from B & O's presidency on June 1, 1941 at age 80. His successor, Roy White, was not one to take the company in radically new directions or, for that matter, to add much to Willard's accomplishments. Under White the Royal Blue Line remained essentially in the Willard mold, with only minor tinkering and without any serious questioning of its economics. While profitability of the New York services probably was never measured, it undoubtedly was dubious. After a brief surge of traffic during the 1939 New York World's Fair, New York passenger traffic totaled only 571,000 people in 1940 — an overall average of 98 passengers per train.

One of White's early moves was to remove the *Columbian* heavyweight streamliner from the New York service in September 1941. Long-distance coach travel, traditionally a railroad stepchild, had been recognized as a promising and profitable passenger market. At the time both the Pennsylvania and New York Central had introduced all-coach luxury streamliners on their New York-Chicago routes which had proved highly popular. B & O decided to convert its *Columbian* into a similar overnight all-coach run between Washington and Chicago, rebuilding its existing equipment and adding more streamlined heavyweights. Still carrying the *Columbian* name, it ran in that

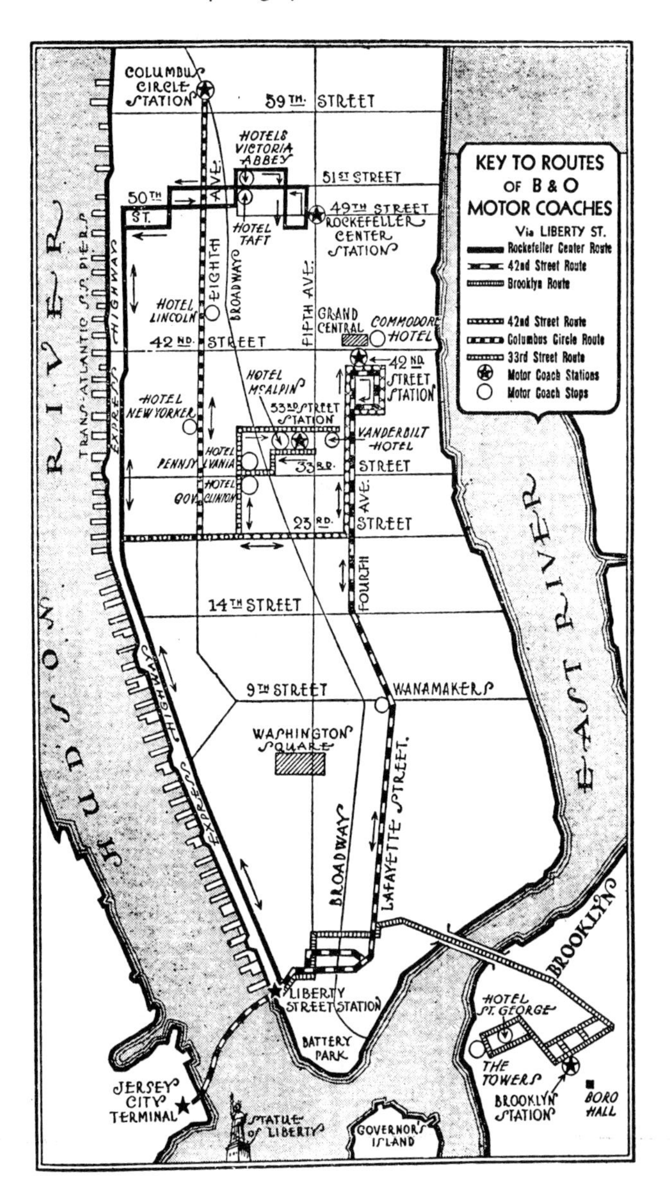

Heading for their train connection, B & O motor coaches ride the Jersey Central's Bound Brook to Jersey City. R. J. Lewis Collection, Francis Palmer photograph.

In another world from the main line, the Landenberg Branch plodded along much as it had since the 1870s. B-18 No. 2012 has the local freight at Ashland, Delaware in 1947. Ten-wheelers such as this often handled local freight work on the Philadelphia line. C. A. Brown photograph.

service until replaced by a new lightweight train (B & O's first and only completely lightweight train set) in 1949. Conventional cars were substituted on the New York run, leaving the *Royal Blue* as the only "pure" New York-Washington streamliner. As noted earlier, heavyweight streamlined cars from trains like the *Capitol Limited, National Limited,* and the reassigned *Columbian* typically were mixed with ordinary heavyweights between New York and Washington.

A much less heralded prewar casualty was one of the two surviving passenger locals on the Baltimore-Philadelphia line. In October 1941 the gas-electric commuter train between Philadelphia and Singerly, Maryland expired, leaving most of the little stations along that section of the line without passenger service for the first time since 1886. Thanks in part to a wartime reprieve and a slightly healthier patronage, the Baltimore-Singerly commuter local hung on for eight more years.

White was also forced to resolve a nasty problem in New Jersey by giving up the B & O's Philadelphia-Jersey City passenger trackage rights which Willard had arranged in 1926. The State of New Jersey had become a notorious taxer of railroad property; in fact, its railroad tax rates were claimed to be the country's highest, and were a factor in the Jersey Central's 1939 bankruptcy. Under the 1926 trackage rights agreement, B & O was considered an operating railroad in New Jersey and thus taxable too. So White had to devise a way of avoiding the liability while still maintaining B & O's control over its New York passenger services.

White's answer was to substitute a complex division-of-revenue traffic agreement for the Reading-CNJ trackage

Photos this page: *Q-4 class 2-8-2s were staple power on through freights during the 1930s and 1940s.* ***Above:*** *No. 4450 brings perishables east through Collingdale, Pennsylvania in 1940. D. A. Somerville photograph.*

And a year later than in the top photo, the 4625 was seen returning empty stock cars west at Rosedale, Maryland. M. A. Davis photograph.

A pair of Q-4s work a long coal train past the one-time suburban station at Holmes, Pennsylvania. C. A. Brown photograph.

rights contract. Technically the new arrangement was close to the original 1886 agreement, where each railroad was responsible for its own part of the operation and shared the revenues. But in this case it really remained a B & O service. B & O continued to supply all the locomotives and cars (which the Reading and Jersey Central technically "rented" while on their lines), set the schedules, and reimbursed its two partners for any losses or extra expenses. Crews of all three railroads were pooled, but as before they wore B & O uniforms. To all outward appearances the trains were still B & O trains and seemed to be running to Jersey City over trackage rights. But as of December 31, 1941, the new agreement took effect, the trackage rights ceased, and B & O was no longer a New Jersey railroad. (The wholly-owned Staten Island Rapid Transit still was, of course, but its New Jersey trackage was negligible.)

By the end of 1941, however, the traffic picture had abruptly changed again. For the next four years the principal problem would not be the Pennsylvania, but B & O's ability to handle a huge increase in business.

RISE, FALL, AND RISE ... PERHAPS

1941-1990

With World War II under way in Europe in 1939, Washington's activity increased dramatically; when the United States entered in December 1941, it exploded. Already busy, the Pennsylvania's New York-Washington line took the brunt, becoming familiarly and unlovingly known as "the big red subway." In the most spectacular performance of their long careers, the Pennsy's GG-1s shuttled strings of twenty or more crowded Tuscan red P-70 coaches and parlor cars along the corridor at 80 mph on hourly headways — often with extra sections in between. Word quickly got around among seasoned "subway" riders that "you can usually get a seat on the B & O." B & O's genteel trains began filling up too, and necessarily lost some of their gentility — although not as much as the Pennsy. The four-hour *Royal Blue* was slowed to four and one half hours and lost its lunch counter-coach and lounge-observation car in favor of more coach seating. And in the war's closing months an Office of Defense Transportation order banning short-haul Pullman travel removed the sleepers from the overnight trains. Dropped in July 1945, the New York-Washington and New York-Baltimore sleepers reappeared March 15, 1946. The lonely Newark-Elizabeth, New Jersey connecting bus service got an immediate execution; it ceased operating in January 1942 and was never resumed.

While Washington was the prime passenger magnet, B & O's Philadelphia and New York trains also handled a

At Philadelphia, an E-6-powered train loads Baltimore and Washington passengers under the 1887 trainshed. B & O Museum Collection, B & O photograph.

Shorn of its green and gold paint and its presidential name, rebuilt P-7 5312 — now in solid blue — takes the Capitol Limited out of Jersey City in 1946. Ted Gay photograph.

heavy intermediate business at on-line military bases. The Army's Aberdeen Proving Ground in Maryland, established in 1917, boomed, both figuratively and literally. Although its property bordered the Pennsy main line and Army-operated shuttle trains ran to the PRR station, Aberdeen soldiers often used the B & O station several blocks west where all trains but the *Royal Blue* would stop on request. Another World War I holdover was Fort George G. Meade, midway between Baltimore and Washington. A regular stream of troop trains moved in and out of the base itself, while hordes of other soldiers were shuttled between Camden station and Fort Meade by buses of the Baltimore

A pair of weary 1912-1913 vintage Q-1s take a heavy Philadelphia freight through Bay View, Baltimore in 1947. B & O Historical Society Collection, W. R. Hicks photograph.

Photos at left and below: *Troop trains also were regular fixtures on both the B & O and Staten Island Rapid Transit.* ***Left:*** *At Ridley Park, Pennsylvania an assortment of venerable Pullmans behind a Q-4 heads east for an embarkation point in 1944. C. A. Brown photograph.*

& Annapolis Railroad and B & O's own West Virginia Transportation Company. In 1942 the Navy took over a one-time private boy's school located on a bluff overlooking the Susquehanna River west of B & O's Aiken, Maryland station. Commissioned in October 1942 as the Bainbridge Naval Training Center, the facility trained over a million recruits before its deactivation in 1947. (Bainbridge was temporarily reopened in 1951 during the Korean war.) The once-quiet Aiken station, in the past merely a local stop, was regularly flooded with blue and white uniforms.

At the top end of the scale were the regular presidential specials between Washington and Jersey City. President Franklin Roosevelt took frequent weekend trips to his family home at Hyde Park, New York on the Hudson River near Poughkeepsie. Roosevelt emphatically preferred rail travel over air, and almost invariably chose the B & O over the Pennsy because, it was said, of his admiration for Daniel Willard and/or his distaste for W. W. Atterbury's Republican activism. Whatever the reasons, the Royal Blue Line

Slightly less impressive but equally purposeful was the SIRT's newly-delivered Alco S-2 No. 488 at Elm Park. Another diesel pushed at the rear. I. Cusick photograph.

was regularly tied up by the prestigious trains which followed the regular B & O-Reading-Jersey Central route to Claremont, outside Jersey City, where they were handed over to the New York Central for the run up the West Shore

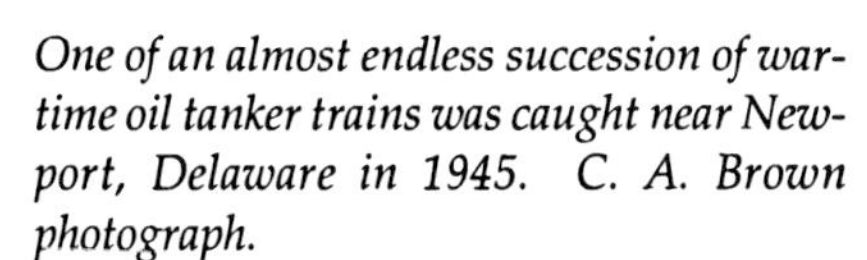

One of an almost endless succession of wartime oil tanker trains was caught near Newport, Delaware in 1945. C. A. Brown photograph.

Since the early 1900s, Staten Island had been the domain of anthracite-burning camelbacks such as B & O's E-13b No. 1633, shown at Clifton in 1939. Ted Gay photograph.

line to Highland, New York. They were undoubtedly a mixed blessing; all opposing traffic had to be stopped, and Roosevelt insisted on a leisurely 50 mph speed limit, probably because lurches and vibration aggravated his physical limitations. In the early years of the war, when the train was handled at Union Station in Washington, the return trips occasionally were held several hours at Fort Meade Junction in order not to arrive too early in the morning at the station.

Between 1940 and 1946 passenger traffic on B & O's New York line leaped an astonishing 104 percent to 1.2 million passengers a year. Unquestionably it was a boon,

Succeeding the steamers for the SIRT's low-speed freight-lugging work was a small roster of utilitarian Alco S-2s. No. 485 tows a transfer drag along the north shore of Staten Island at Sailor's Snug Harbor. I. Cusick photograph.

To keep train movements fluid, pairs of passing sidings were spaced roughly ten miles apart along the Philadelphia line. The westbound Royal Blue overtakes a freight at Holmes, Pennsylvania in 1948. C. A. Brown photograph.

but also a revealing commentary on how under-used the route had been before.

Freight flooded the Philadelphia-New York route too, particularly military export traffic for New York Harbor and solid trains of petroleum from the Gulf Coast, forced onto the rails by German U-boats along the Atlantic coast.

The little Staten Island Rapid Transit suddenly became strategically important and sagged under the load. Freight tonnage, which had been swelling since the European war began in 1939, more than doubled between 1942 and 1944, hitting a record 3.2 million tons that year. In addition, piers at Stapleton, just south of St. George, were used as a major Port of Embarkation, handling over 742,000 troops and receiving 100,000 prisoners of war. Halloran Hospital on Staten Island was made a rehabilitation center for returning wounded, which afterwards were returned to their homes on "ambulance trains" which the SIRT moved out to the PRR, Lehigh Valley, B & O, Reading, and Jersey Central. At times up to 75 military hospital cars were held at the SIRT's Arlington yard.

The wartime crush also finally brought about the SIRT's dieselization. Between 1943 and 1944, eight Alco model S-2 1000 hp switchers and a GE 65-ton 400 hp centercab arrived to aid the struggling 0-6-0 and 2-8-0 camelbacks, which included several weary B & O E-13s and D-23s. Painted in B & O's standard blue and gold diesel switcher scheme and numbered in B & O's system series, they were nonetheless owned by the SIRT and lettered "Staten Island." With the traffic dropoff in 1946 the Staten Island retired its last steam power and the nine diesels took over all freight work, supplemented where needed by B & O power. The SIRT occasionally reciprocated, sending its GE centercab to B & O's 26th Street freight terminal when the old boxcab was in the shop.

The postwar letdown came, of course, although not abruptly. Between 1946 and 1949 passenger traffic on B & O's New York trains dropped 33 percent, but then briefly stabilized at a level higher than the prewar years. Helping sustain it was a healthy tour group business; indeed, fully half of B & O's New York traffic consisted of various types of private tour groups or organizational outings which took advantage of B & O's "private" bus service to and from their New York hotels.

Except for a brief service cut between early 1945 and April 1947, B & O continued its traditional eight New York-Washington round trips, now supplemented by two Philadelphia-Washington runs each way. In addition, solid mail-express trains Nos. 29-32 traveled the New York-Washington line at night carrying a heavy business between B & O's East Coast cities and various cities on the Chicago and St. Louis lines. Probably the longest regular first-class trains on the New York line, Nos. 29-32 hauled up to eighteen head-end cars, plus a rider coach and caboose.

With the war's end the *Royal Blue* got back its lunch counter car, lounge-observation car, and four-hour schedule. One other train, the New York-Washington late-afternoon *Marylander* (now the *Congressional's* competition) also made the trip in four hours. In August 1947 the *Royal Blue*, along with the *Congressional* and three other premier PRR and New York Central trains, got public radio telephone service — the first such regular operation in the United States.

A late afternoon departure from Jersey City, the New York-St. Louis Diplomat crosses the Jersey Central's long Newark Bay bridge in 1954. H. H. Harwood, Jr. photograph.

As before, the eight through New York-Washington trains performed a variety of functions. Three of them, the *Royal Blue*, the *Marylander*, and the *Night Express*, essentially were strictly New York-Washington runs. Three others, the *Capitol Limited*, the *Shenandoah*, and Nos. 9-10, carried through cars over the New York-Washington-Pittsburgh-Chicago route in addition to New York-Washington coaches and diners. The *National Limited* and *Diplomat* did the same for New York-Washington-Cincinnati-St. Louis business. Although these trains carried the same names and numbers all the way between New York and their western terminals, they actually were different trains west of Washington; the through cars were switched in and out at Union Station. Technically they competed with the Pennsylvania and New York Central between New York and Chicago and St. Louis, but in fact everyone knew they did not. The all-Pullman *Capitol*, for instance, took 21 hours 25 minutes to make the trip and the all-coach *Columbian* followed ten minutes later. Both the Pennsy's *Broadway Limited* and the Central's *20th Century Limited* were scheduled for sixteen hours, and their all-coach *Trail Blazer* and *Pacemaker* for seventeen hours. B & O's *National Limited* spent 24 hours 55 minutes traveling between New York and St. Louis; the Pennsy's all-coach *Jeffersonian* did it in twenty hours 25 minutes, and the Central's *Southwestern Limited* took 21 hours 55 minutes. The B & O trains, however, did do a moderately healthy business at intermediate points.

During this Indian Summer of railroad passenger service, some lightweight sleepers and the new *Columbian* coaches also appeared on the New York trains, and through sleeper services were operated to a fascinatingly wide (and ever-changing) variety of destinations. At various times

Scheduled for Philadelphia-Washington "commuters," train No. 35 (later No. 21) started its run at Wayne Junction in Philadelphia in the early morning. With three head-end cars, a single coach, and a parlor-diner-lounge car, it passes the 1886 station at Holmes, Pennsylvania in 1948. C. A. Brown photograph.

The postwar years brought gradual dieselization all along the Royal Blue Line route. In this 1954 scene at Jersey City terminal, an inbound Jersey Central Alco RS-3 passes an aristocratic set of B & O E-units waiting to leave for Washington. H. H. Harwood, Jr. photograph.

during the late 1940s and early 1950s, for example, the eastbound *National Limited* carried sleepers to New York from St. Louis, Louisville, Morgantown (West Virginia), Parkersburg (West Virginia), Wheeling, Pittsburgh, Oklahoma City, and Tulsa. Between June 1946 and January 1947, it also carried a through New York coach from Fort Worth, Texas in addition to its normal St. Louis coach. The westbound *Diplomat* returned many of these exotic cars from New York.

The year 1948 saw the start of one of those peculiar folksy traditions at which the B & O seemed to excel. Just east of the Philadelphia line's station at Jackson, Maryland (about two and one half miles east of the Susquehanna River), a huge old holly tree grew next to the B & O tracks. The B & O purchased the property back in 1930 to preserve the tree, and left it in its natural state. But beginning in 1948, the "B & O Holly Tree" became a corporate institution. Every Christmas season it was elaborately decorated and lighted, and special trains were run from Baltimore for the lighting ceremony — often accompanied by the B & O Glee Club — which was broadcast on a Baltimore radio station. Regular passenger trains were slowed while passing it, car lights were dimmed, and trainmen went through the cars announcing (from a standard script) "Watch for B & O's beautiful holly tree on the right (or left) — in just a few minutes." Throughout the B & O system, passenger trainmen wore holly sprigs in their uniform lapels for the season. The tree also was included in the company's official list of stations and sidings. Somehow, Daniel Willard's ghost was still active in B & O's affairs.

More passenger diesels had appeared on the New York line in the 1940s to replace the workhorse P-7 Pacifics, which in turn began to filter out to B & O runs elsewhere. An order of EMD model E-6s had arrived in 1940 and E-7s were delivered in 1946. While assigned system-wide, as many as possible were put on New York trains. On September 28, 1947 B & O announced that all its regularly-scheduled Washington-Jersey City passenger trains were diesel-hauled. Steam was far from dead, however; it still worked the Philadelphia-Washington expresses, most special passenger trains and troop trains, and most freights.

Aside from this, B & O began seeing the handwriting on the wall and avoided making any major investments in new Royal Blue Line passenger equipment. Although some postwar lightweight equipment had begun to show up on the through western runs, the New York trains typically were a mishmash of prewar streamlined heavyweights, conventional heavyweight cars, and a limited scattering of new ones. Car colors were equally mixed: some blue and gray, some solid blue, and a few still olive green. Significantly, no move was made to modernize the line's premier *Royal Blue*. Somewhat stolid and stodgy even when new, the "Blue" remained the 1937 train, sometimes supplemented with a conventional heavyweight coach. The Pennsylvania's supremacy in the New York market was demonstrated all too vividly in 1952 when it reequipped its *Congressional* with all-new Budd-built stainless steel cars. In contrast to the weary eight-car *Royal Blue*, each of two *Congressional* train sets totaled eighteen cars, including eight coaches, seven parlor cars, a full lunch counter car, and a twin-unit diner, and ran on a three hour 35 minute schedule.

B & O doggedly kept trying, although now only halfheartedly. The aggressive advertising of the Willard

Another new White rides the unique turntable at the west end of the Jersey City terminal driveway. I. Cusick photograph

era was muted, but money was spent to modernize the distinctive (but now worn-out) 1936 New York bus fleet. Between 1949 and 1951 it phased in 30 new White model 1136s, receiving ten a year. While specially styled and painted, they closely resembled conventional suburban buses of the period and never quite matched the appeal and distinction of their predecessors.

At the same time, however, more services were cut. The Baltimore-Singerly "motor" train, the last all-stops local on the Philadelphia line, ended in May 1949. More significantly, one New York-Washington and one Philadelphia-Washington round trip disappeared from the timetable April 29, 1950. Afterwards only seven New York trips remained, plus one Philadelphia run and one solid mail-express train each way.

The freight picture brightened a bit. Always woefully weak in on-line industry, the Philadelphia line finally got a large plum in 1946 when General Motors was enticed to build a new automobile assembly plant adjacent to Wilsmere Yard outside Wilmington. Production began in the fall of 1947 and grew to the point in 1955 where 4400 employees were producing sixty cars an hour on two shifts. It remained the line's largest single industrial customer through the 1980s.

At the New York end a new harborfront coal dumper was built in 1949 at Howland Hook, on the northwest corner of Staten Island near Arlington. (The old facilities had been shut down in the early 1930s.) Less than ten years later, New York's Consolidated Edison Company completed a large coal-fired generating plant on Staten Island at Travis and became a heavy B & O coal consumer.

Almost at the same time, the SIRT got a world-class engineering landmark. The now-ancient 1889 Arthur Kill swing bridge increasingly had become a menace to the large ships now using the channel. By then the New Jersey side of the Arthur Kill had become almost solidly built up with oil refineries and chemical plants. In November 1947 an errant oil tanker had hit the bridge, forcing all Staten Island freight to be rerouted through Jersey City and floated across to St. George. Additionally, the old bridge hampered B & O's new Staten Island coal traffic, since it was limited to fifty-ton hoppers. In the early 1950s B & O and the federal government jointly agreed to replace it with a long-span vertical lift bridge — and indeed the 558-foot lift span was to be the world's longest, snatching the record from the 544-foot Cape Cod Canal bridge. (In 1990 it still

The third generation of B & O "motor coaches" was the most ordinary. A pair of the new Whites unloads a group of high school students returning from a class trip to New York. Catering to special tour groups such as this was always a B & O specialty in New York, and generated over half of the railroad's New York passenger traffic. W. D. Edson photograph.

held that record.) Work on the new Arthur Kill lift bridge began in 1955 and it was opened in August 1959.

Elsewhere in New York Harbor, merchandise traffic continued to be healthy through the late 1940s and 1950s, particularly inbound perishables and outbound carloads of small shipments from freight forwarders. By then freight forwarder traffic was by far the single largest outbound "commodity," and various forwarder companies leased most of B & O's pier station facilities for consolidation and loading. But if the volume was healthy, the profitability probably was not. Virtually all the inbound perishables and outbound forwarder business was carfloated between Jersey City and Manhattan, where it was loaded or unloaded at expensively leased piers. Even costlier were the lighterage shipments, often necessary for export and import traffic, since these required manual reloading between freight car and lighter. All this carfloating and lighterage was no small operation either. In 1949 B & O's marine fleet alone (not including partner CNJ) included seven tugs, 21 carfloats, five steam lighters, thirteen refrigerator lighters, and 51 regular lighters. Yet B & O and most other New York railroads performed all these services free to their customers. In part they had to, in order to compete, since the New York Central could offer direct rail service in Manhattan.

The germ of a solution appeared in 1954 when B & O began a trailer-on-flatcar service (universally called

Throughout the steam era, Reading and CNJ power usually handled B & O freights east of Philadelphia. Here a heavy Jersey Central 2-8-2 comes off the B & O at Park Junction heading for Jersey City.

Similar power with a similar train crosses the Schuylkill River on the Reading's Columbia bridge at Belmont. Both photos date to 1951. W. P. Ellis photographs.

Opened in 1959, the SIRT's Arthur Kill vertical lift bridge was, and still is, the world's longest. R. J. Cook photograph.

"TOFC" or "piggyback") on its system, one of the first railroads to do so. The start was very modest. Initially, Philadelphia was as far east as B & O could run its TOFC shipments; the Jersey Central, its finances once again wobbly, balked at building a Jersey City piggyback terminal because of the cost and its own small share of the revenues. Nor was B & O's Philadelphia terminal suited for much volume. Typical of early TOFC facilities, it consisted of a small stopgap ramp terminal at B & O's team track yard along the Schuylkill River at 25th and Locust Streets. Initial piggyback service between Philadelphia, Pittsburgh, and Chicago started July 20, 1954 and gradually was extended to other western terminals in 1955. After agreeing to build its own terminal on Jersey Central's land, B & O was able to start Jersey City service in 1957.

In theory at least, TOFC (which B & O christened "Tofcee") was an answer to the two knottiest problems of B & O's Philadelphia and New York freight operations. Not dependent on rail sidings or water transportation, piggyback trailers could reach customers off B & O tracks (and most customers were) and also could be driven directly to overseas shipping piers or inland customers in New York. But it would be at least ten years more before rail-highway intermodal services matured enough to make a difference.

Dieselization continued into the early 1950s and was concluded by early 1954 on the Washington-Baltimore-Philadelphia lines. New EMD E-8 passenger units arrived in 1950 and 1953, and a steady stream of freighters, EMD F-3s and F-7s and Alco FA/FBs, poured in between 1948 and 1953. In 1952 the pioneering Baltimore Belt Line electrification was shut down. Although some steam still remained then, the few steam-powered trains were hauled through the tunnel by a pair of newly-delivered Fairbanks-

This photo and photos on opposite page: *Picturesque but hideously costly, the jointly-operated B & O-CNJ New York floating operations struggled on through the 1950s and 1960s.* ***Left:*** *The diesel tug Howard E. Simpson noses a lighter-barge tow into St. George in 1954. I. Cusick photograph.*

Photos this page: *A pair of carfloats negotiates the East River after working the Long Island Railroad interchange. H. H. Harwood, Jr. photographs.*

Morse 1600 hp road-switchers supplemented where needed by 1000 hp and 1200 hp F-M yard switchers.

Roy White ended his rather lackluster B & O presidency in September 1953. His odd choice of successor, Howard Simpson, originally had come from the Jersey Central and had spent most of his career in passenger sales work; he was, not surprisingly, a passionate Royal Blue Line booster. Unhappily, Simpson was forced to become its executioner.

The New York passenger picture had become genuinely grim and was worsening steadily. By 1952 passenger loadings had taken another clear downturn, this time with no sign of recovery. Between 1951 and 1955 they dropped thirteen percent, and were a disastrous 27 percent below the 1947 level. Simpson tried two palliatives in 1956. One more New York-Washington round trip, the *Marylander*, was removed in late October, leaving six New York passenger trains and the mail-express run each way.

At the same time, radical surgery was performed on the single surviving Philadelphia-Washington run. Its schedule was combined with the semi-local Baltimore-Washington-Pittsburgh-Cleveland *Washingtonian* and converted to a fast daytime Philadelphia-Pittsburgh motor train service. Christened the *Daylight Speedliner*, the new train consisted of a three-car set of high-speed self-powered Budd RDC cars operating in each direction. The new RDC train was loosely patterned on European intercity motor train services of the time and included reclining coach seats and a small dining section, the only American RDC "diners." Following the traditional pattern, the *Speedliner* left Philadelphia in early morning to carry the Washington "commuters" before heading west to Pittsburgh; its eastbound mate returned to Philadelphia in late afternoon. Another creative experiment, the *Daylight Speedliner* was to

On its way to New York, the Capitol Limited drums past Waverly Tower in Baltimore, once the east end of the Belt Line electrification. It is 1953 and the center "pen" track is empty and the third rail now gone. R. L. Wilcox photograph.

be the last chapter in the Royal Blue Line's distinguished history of passenger service innovation.

Nothing helped, of course; the curve continued relentlessly down. By 1957 New York-Washington passenger loads were half what they had been in 1946 and over 34 percent below the more normal year of 1947. The six surviving round trips were now averaging 92.5 passengers per train — less than two full coach loads. Certain specific periods were even worse. In early November 1957 some individual trains averaged as low as seventeen paying passengers per train. But the fatal problem was the New York bus operation. Always extraordinarily expensive, it became much more so as loads diminished and the high fixed costs were spread over fewer passengers. Simpson resolutely maintained the traditional five motor coach routes and terminals, even though the 42nd Street service was the only one with a significantly heavy business. (In the last years the Brooklyn route often carried only two or three passengers per trip.) B & O's cost for carrying each New York bus passenger worked out to $4.69, while the price of a New York-Baltimore ticket, which included the bus service, was only $6.87. It was, in a word, hopeless.

It also was an all-or-nothing situation. Although all railroads were suffering from ever-diminishing passenger traffic, they usually were forced to follow an attrition strategy, cutting off a few trains at a time but still maintaining some basic services over their major routes. But in this case, the heavy fixed costs of the Jersey City terminal and the New York bus operations and stations, coupled with the

Posed at Philadelphia's 24th and Chestnut Streets station, the shiny new Daylight Speedliner worked less than two years on this end of its run. As of 1990 virtually all of its equipment still existed, some of it working. H. H. Harwood, Jr. Collection, B & O photograph.

Photos this page: *Passenger traffic may have been withering, but the Philadelphia line still saw some heavy special trains.* ***Above:*** *On a cold late November day in 1952, a pair of P-7s hurries an Army-Navy football special past Cowenton (now White Marsh), Maryland.*

Warmer weather brought racetrack trains. In 1955 three F-units are heading for Delaware Park at Poplar, Maryland, near Rossville. M. A. Davis photographs.

The once-immaculate passenger service was beginning to look moth-eaten in 1957 as a westbound pulls into the deserted station at Chester, Pennsylvania. The building's clock tower and platform canopy have been removed. This location is now a monoxide-filled gulch occupied by Interstate 95 and a single railroad track. Ken Douglas photograph.

Pennsylvania's clear dominance of the market and extensive duplicating services, dictated complete abandonment.

Late in 1957 B & O filed the official petitions to end all passenger service east of Baltimore. Ever the effective salesman, Howard Simpson personally traveled the circuit of on-line cities explaining the B & O's problems. Thanks largely to his efforts there was no opposition — something unheard-of in passenger abandonment cases at the time and especially amazing, considering it involved all service on a major line.

Saturday, April 26, 1958 was set as the last day. The full day's schedule was operated; indeed, the traditional tour group business was solicited and handled to the end. On that final Saturday, two special trains and no less than seventeen smaller tour groups were handled through Jersey City, most of them schools taking their spring trips. Inevitably, too, several carloads of railroad enthusiasts from Baltimore and Washington took their last round trips to New York.

Train No. 8 was the final eastbound run, arriving at Jersey City at 9:15 p.m. Westbound service actually ended the next day when the overnight train, No. 11, completed its trip early Sunday morning. To help clear out the Jersey City yard, No. 11's normal ten-car consist had swelled to 21 cars, including ten coaches and four diners. Ahead of No.

During the last week of passenger service, a pair of E-8s waits with the usual assortment of heavyweight cars at Elizabeth, New Jersey. By the 1980s, the one-time Jersey Central four-track main line through here had been reduced to a single-track Conrail industrial branch. R. J. Yanosey Collection.

11 had come a sad procession: first an F-7-powered extra from Philadelphia with seventeen passenger cars, including three *Daylight Speedliner* RDCs. Following it was No. 29, the mail-express train, which arrived in Baltimore with 31 cars from Jersey City trailing behind its three GP-7/GP-9 diesels — probably a record B & O passenger train length. Baltimoreans probably grieved the most, since they had always viewed the Royal Blue Line trains as their very own, a far more civilized alternative to the Pennsylvania's "red subway." They were right; but sadly, civilized tastes did not translate into heavy patronage.

Baltimore-Washington passenger service continued, and Mt. Royal station remained open as a terminal for a few Baltimore trains including the still-operating *Daylight Speedliner* service to Pittsburgh. The large and now mostly empty station was living on borrowed time, however; it finally closed completely in April 1961. Happily, in 1964 it was saved from demolition by a transfer to the Maryland Institute College of Art. Its interior was altered to accommodate studios and galleries, but Francis Baldwin's Romanesque exterior and the station's now-rare trainshed remain today essentially as they were in 1896.

Most other Royal Blue Line stations were quickly demolished, although Frank Furness's eccentric Philadelphia structure survived until it burned in 1963. As of early 1990, only two of the original 1886-era stations remained: one at Aberdeen, Maryland (probably a Furness building, now used by railroad maintenance forces), the other the derelict Furness-designed Market Street station in Wilmington, located near the present Amtrak station. A third and final survivor is the Newark, Delaware station, a bland 1945 replacement of the picturesque original.

Few railroad main lines had such an abrupt and traumatic change as the B & O's Baltimore-Philadelphia line. But as it turned out, this would be merely the start of a slow transformation which would completely alter its physical character, its operations, and even its basic mission.

The physical change came almost immediately. Besides the almost complete removal of passenger stations, there seemed to be no need for a double-track railroad with its regularly-spaced extra passing tracks and interlocking towers. Plans were made to single-track the route, installing Centralized Traffic Control and eliminating virtually all intermediate interlocking towers. Single track began at Rossville, Maryland (four miles east of Bay View yard) and ended at Darby, Pennsylvania (west of Philadelphia). In between, five long passing sidings, spaced roughly ten to twelve miles apart, were installed at Van Bibber, Aiken, and Singerly in Maryland, at Wilsmere yard in Wilmington, and at Feltonville, Pennsylvania. The Howard Street tunnel and most of the Baltimore Belt Line also were single-tracked. The project was completed in 1960, and John Garrett's railroad emerged as lean and austere with, it seemed, few trains.

The "few trains" was partly an illusion, and therein lay a problem. When designing the single track layout, the planners apparently did not carefully analyze the line's traffic patterns. Although the end of passenger service significantly reduced total train movements, the passenger trains had run mostly in the daytime while the freights were bunched at night. In 1968, for example, there were thirteen manifest freight trains scheduled to operate over the line between 11 p.m. and 7 a.m., and the inevitable deviations and lapses from the printed schedules could create even worse concentrations. As a result, a major peak load still remained, and the line suffered serious congestion. Throughout the 1960s and early 1970s, occasional proposals were made to add more second track, although none was carried out.

East of Baltimore, single track started at Rossville, Maryland, shown here in the CSX era. In this 1989 scene a GP-35-powered eastbound local waits for train 377 to clear. H. H. Harwood, Jr. photograph.

The demise of passenger service, along with single-tracking, brought the Philadelphia line closer to economic reality after so many years of essentially artifical support. But some even greater problems remained to be solved. In 1961 B & O again changed its executive management, and in early 1963 it also came under the Chesapeake & Ohio's control. The two events produced many changes in corporate direction and marketing philosophy. Jervis Langdon, Jr., B & O's new president, was quick to face the true economics of the New York freight operations, and for the first time company officials began debating whether it was even worth staying in the New York market. The traditional waterborne operations, switching, general congestion, and obsolete facilities at the port all made the business almost hopelessly unprofitable. In fact, the high switching charges on some Brooklyn traffic produced a direct loss before the shipment even moved over B & O rails. The only major exception was coal handled over the dumpers at Jersey City and Howland Hook.

Another concern was the financial fortunes of the Jersey Central and Reading. The CNJ was especially precarious. Having emerged from a ten-year bankruptcy in 1949, it now seemed headed toward another. Essentially a short-haul carrier, the Jersey Central was cursed by extensive terminal operations in a notoriously high-cost area. Its parent, the Reading, was somewhat healthier, but only by comparison. Both the CNJ and Reading had lost most of the anthracite coal business which traditionally had supported them, and they were seriously affected by the traffic malaise and cost problems of the northeastern railroads. Both companies also suffered a heavy financial drain from their extensive commuter service operations. (At this time public policy was only cautiously inching toward subsidy and ownership of suburban lines.) As the solvency of the two railroads became more questionable, their interests and the B & O's became more divergent. For example, the Jersey Central and Reading earned more on freight from New York to the west by routing it through Allentown, Harrisburg, and Shippensburg, Pennsylvania, than by carrying it to Philadelphia for the B & O.

Increasingly the B & O found itself in a quandary. Although it had effective control of the Reading, and of the CNJ through the Reading, it still owned less than fifty percent of the Reading's stock. Thus it had to be cautious in its corporate relationships; there was always the danger of minority stockholder objections if the B & O forced the Reading to do anything financially detrimental to it. Yet B & O was not in a position to support either company or absorb them itself. It too was financially weak and, possibly more important, its new C & O parent had other priorities. About 1963 B & O had an opportunity to boost its Reading holdings to 70-75 percent. Although Langdon supported the purchase, C & O turned it down, ostensibly because of the dubious long-range New York outlook.

Although gloomy, the Reading's situation at least seemed salvageable. Former B & O operating vice president Charles Bertrand was sent to the Reading as executive vice president in 1963 and was made president the following year. Bertrand began to turn the company around, but lack of support from both C & O and the New York Central (which at that time still had its minority interest) helped to doom his cause. Both Bertrand and his CNJ counterpart, Perry Shoemaker, felt increasing pressures to protect the interests of their own companies. Disputes arose with B & O over revenue divisions, service requirements, facilities costs, and freight routings. As the New York terminal operator, the beleaguered Jersey Central became particularly confrontational, and the traditional "happy family" atmosphere of the B & O-Reading-CNJ relationship began to vaporize. In March 1967 the Jersey Central entered still another bankruptcy — one from which it would not emerge. Reading followed in November 1971, and in mid-1973 B & O liquidated its Reading stock ownership. The stage now was set for the possible loss of B & O's New York entryway.

Despite the long-range problems, Langdon and his C & O successors worked to maintain the New York and Philadelphia freight services and make them more viable. To overcome B & O's chronic inability to reach Philadelphia industry, Langdon negotiated a special agreement with the Reading in late 1963. Freight customers in the area served by the Reading were opened to B & O routings through Philadelphia; in exchange, B & O protected the Reading's revenue by paying it the full amount the Reading would have received if the shipment had moved over its long haul to Shippensburg, Pennsylvania. The Reading continued to switch the customers itself, interchanging the cars with B & O through Park Junction.

But the most significant boost in business over the B & O-Reading-CNJ route came from the large increase in bituminous coal traffic. A strong proponent of innovative pricing techniques, Langdon designed low unit train rates for coal moving to eastern receivers such as Consolidated Edison in New York, Philadelphia Electric, and some New Jersey utilities. At first the B & O coal was routed to the Reading through Shippensburg, but after a similar agreement to protect the Reading and Jersey Central's long-haul revenues, it was switched to Park Junction. In getting this business for the B & O, Langdon managed to outmaneuver the Pennsylvania; more importantly, however, his systemwide unit train rates helped reverse the B & O's financial fortunes in 1962 and also shortstopped plans to build a coal slurry pipeline to the East Coast.

Beginning in 1962, Langdon also took the lead in what eventually became a general withdrawal of all waterborne freight service in New York harbor. He began giving up the Manhattan pier leases, trimming down carfloat opera-

Typical of the motive power mishmashes of the 1960s was this array, shown leaving Potomac Yard with the Hudson, a runthrough freight for Jersey City. Two CNJ Fairbanks-Morse "Trainmasters" trail a C & O F-7 and B & O GP-9 in February 1965. H. H. Harwood, Jr. photograph.

tions, and encouraging the freight forwarders to truck their traffic to Jersey City for loading there.

At the same time, the Langdon regime put heavy emphasis on rail-truck intermodal operations to bypass harbor operations and also to reach new customers. B & O offered piggyback service for its New York freight forwarder and export-import business, and expanded the Jersey City TOFC terminal; in 1972 it was relocated to Elizabethport, New Jersey adjacent to the new containership terminals at Port Newark and Port Elizabeth. A more modest rail-truck bulk service was also started at Jersey City, primarily for inbound chemicals. As before, CNJ continued to operate all of the New Jersey intermodal facilities. The Philadelphia TOFC terminal was moved to Delaware Avenue and Jackson Street near the waterfront, a more spacious area although awkward to reach by rail.

And in 1964, for the first time in Royal Blue Line history, the B & O, Reading, and Jersey Central began pooling locomotives on through freight runs. A pair of runthrough schedules was established between Jersey City and Potomac Yard at Alexandria, Virginia, and for the next several years railfans were delighted by random mixtures of diesels from all three railroads plus C & O. The Jersey Central's contributions often included Alco RS-3s and its rare Fairbanks-Morse Trainmasters and, in the late 1960s, Norfolk & Western and B & O F-7s which had been leased to the ailing line. Although operations later changed, B & O/C & O power continued running through to Jersey City or Elizabethport until the early Conrail era.

By the early 1970s the entire makeup of New York freight traffic was changing rapidly. All of the area railroads had begun phasing out their water operations, and at the same time the traditional carload export-import business moved to containers. In 1971 B & O stopped all remaining free lighterage services. Two years later the Jersey Central gave up all its marine operations, essentially abandoned Jersey City as a terminal, and pulled its operations back to Elizabethport on the west side of Newark Bay. B & O held on slightly longer, moving its carfloat terminal back to St. George and continuing some token floating for freight forwarders located on piers at West 23rd Street and West 26th Street. The Brooklyn Eastern District Terminal took over some other marine services. On September 20, 1976, B & O finally shut down too, turning over the St. George operation and some remaining equipment to the New York Dock Railway, which by then had absorbed most of the surviving Brooklyn terminal switching properties. The last large railroad to enter New York Harbor, B & O was also one of the last to leave. New York Dock continued using St. George until 1980 when the terminal was abandoned.

The Staten Island operation also significantly changed, mostly for the worse. The always-unwanted passenger service had long since become a clear financial albatross, and in 1953 the SIRT made the first major move by eliminating all passenger service on its lightly-used South Beach and Arlington branches. The South Beach line was abandoned outright; Arlington, of course, continued to carry freight between St. George and New Jersey. The following year, thirty of the electric passenger cars and five

B & O diesels also became regular visitors to the Jersey Central main line, operating into the early years of the Conrail era. A pair of GP-40s brings an eastbound freight past the high-level platforms of the now-demolished Elizabethport station in 1967. R. J. Yanosey photograph.

trailers were sold to the New York City Transit Authority for service on the BMT lines. So, thirty years late, at least some of the Staten Island cars ran in the New York subway system. Beginning in 1959 New York City subsidized the remaining rapid transit services on the St. George-Tottenville line, and in 1971 B & O sold this section to the city. B & O retained the freight line between St. George and Cranford Junction as well as freight-only trackage rights on the Tottenville line, and reorganized the operation as the Staten Island Railroad Corporation.

Shortly afterward B & O went through a corporate change of heart which helped dry up more business on the Philadelphia line. With carload traffic at New York disappearing and trucks taking more of the Philadelphia perishables business, the line's major hope lay in trailer-on-flatcar business. During and after the Langdon era, TOFC on the Philadelphia line had grown substantially, fed by expedited all-TOFC trains to Chicago and St. Louis. But in the early 1970s a new management concluded that the traffic was only marginally profitable and began de-emphasizing it. While never abandoning the service outright, B & O downgraded it, removing most of the special trains and ceasing to solicit the business aggressively.

Through all this shrinking, the B & O-Reading-CNJ freight route continued to struggle along, although the rapidly disintegrating finances of B & O's two partners began creating service quality problems as well as more traffic reroutes. By then the entire eastern railroad situation seemed hopeless. Penn Central, Erie Lackawanna, Lehigh Valley, Reading, Jersey Central, and the little Lehigh & Hudson River all were in what looked like insoluble bankruptcies, and a government-financed consolidation of these companies was planned. In 1975 the C & O-B & O group, which now called itself the Chessie System Railroads, made a surprising and curious corporate turnabout and proposed to buy large portions of the Reading and the Erie Lackawanna. The purchase would have established Chessie directly, heavily, and permanently in the New York area and greatly strengthened it in Philadelphia. Chessie's optimism did not extend to the Jersey Central, but it planned to get trackage rights over former CNJ lines into New Jersey and also have access to industries along the "chemical coast" between Elizabethport and Perth Amboy. Also, according to its preliminary plan, it would remain in the carfloating business between St. George and Brooklyn. Labor opposition to the terms killed the purchase idea; on April 1, 1976 the Reading and Jersey Central (along with the Erie Lackawanna) disappeared into the newly-created Conrail which was now B & O's major competitor.

The exit from New York was not instantaneous, however. For its first several years, Conrail continued to operate a single connecting train service for B & O traffic between Philadelphia and New Jersey, primarily for Staten Island business. It also maintained the traditional Reading and Jersey Central tariff routings, which at least allowed B & O to solicit interline business where it could. But as expected, Conrail began a regular process of diverting the one-time B & O-Reading-CNJ traffic to its own through routes, and B & O's New York-New Jersey business slowly dried up. The final nail in the coffin came in the early 1980s when Conrail simply canceled the old interline tariff routings for any shipments it could handle directly on its own lines. Since Conrail competed at almost all B & O-served points, its latest program almost completely removed B & O from what remained of any markets east of Philadelphia.

Photos this page: *The new era mixes with the old. Powered by a pastiche of diesels from the RF & P, L & N, and Seaboard System, the westbound Orange Blossom Special (now simply CSXT train 171) crosses the 1909 Susquehanna bridge at Havre de Grace in 1988. H. H. Harwood, Jr. photograph.*

A slow recovery started in the early 1980s as Chessie began devising ways of expanding its northeastern markets with various types of rail-truck intermodal services. In October 1980 it created its own common carrier truck subsidiary, Chessie Motor Express — usually called CMX. Using CMX and a different method of structuring rates and soliciting business, the railroad felt it could make the piggyback business more profitable. CMX based its operations at B & O's Philadelphia TOFC terminal, making it the hub for its trucking services throughout the northeast. Once again TOFC services on the Philadelphia line improved and grew; the expedited trains returned and, in fact, the badly-located Philadelphia terminal rapidly became overcrowded.

Other opportunities opened after the 1980 formation of CSX Corporation, a holding company created to control Chessie and a group of southeastern railroads which later were combined as the Seaboard System. Traditionally, B & O's relatively short 140-mile haul between Philadelphia and the southeastern gateway at Potomac Yard had not given it enough revenue to cover certain high-cost traffic like TOFC over this route. But with a large south-

The last active original Royal Blue Line station at Aberdeen, Maryland stares blankly out at a mixture of power from the New York, Susquehanna & Western, and Norfolk & Western, heading east with a Delaware & Hudson train from Potomac Yard to Binghamton, New York. H. H. Harwood, Jr. photograph.

eastern railroad system now in its family, B & O's own short haul was less relevant and new north-south movements could be developed.

The first and most dramatic of these was the *Orange Blossom Special*, a tightly-scheduled all-TOFC train designed to put the railroad back into the movement of Florida perishables to the northeast — a business long since lost to trucks. Created by Seaboard System president Richard Sanborn, the *Orange Blossom* ran through to Philadelphia, where CMX took the refrigerator trailers over the road for delivery in Philadelphia and the New York-New Jersey area. Except for crew changes, the train was handled non-stop; between Richmond and Philadelphia, diesel locomotives of Chessie and Richmond, Fredericksburg & Potomac were pooled, with the RF & P units usually leading.

The *Orange Blossom Special* started in the fall of 1983 as a seasonal train and steadily grew to the point where the already-crowded Philadelphia terminal had to be supplemented. A temporary satellite terminal was built at Wilsmere Yard while Chessie tried to find a suitable central site to handle all the piggyback business. A plan to rebuild Wilsmere as the primary terminal was frustrated by neighborhood opposition, and Chessie could not find another adequate location between Wilmington and Philadelphia. In the end it decided to live with the Philadelphia waterfront site and expanded it — demolishing part of the 1926 Produce Terminal complex in the process. It was an apt symbol of how the railroad business had changed.

Some pressure was taken off Philadelphia in March 1986 when the *Orange Blossom's* run was extended over Conrail from Philadelphia to Conrail's massive TOFC-container terminal at Kearny, New Jersey, near Jersey City. The interline traffic arrangement with Conrail was a 1980s revival of the B & O-Reading-CNJ agreement of 100 years earlier, with Conrail crews handling the train east of Philadelphia and Conrail receiving part of the revenue. The pooled RF & P-CSX power operated through, however. By this time the train had ceased to carry exclusively Florida perishables and was a heavy year-round general freight/perishables TOFC service.

The new and slightly friendlier Conrail relationship was extended in November 1986 when B & O granted trackage rights for one Conrail freight each way between Philadelphia and Potomac Yard. It was still another symptom of the radically changing northeastern railroad geography. Amtrak had taken over ownership of the old Pennsylvania Railroad-New York-Washington line and, with government financing, had rebuilt it as a 125 mph passenger railroad. Neither Amtrak nor anyone else wanted freight on the line — particularly after the tragic Conrail-Amtrak collision at Chase, Maryland on January 4, 1987. While complete removal of freight trains from the corridor was impossible, the old Royal Blue Line was viewed as forming the future freight main line between Philadelphia and Washington. As of early 1990 the full program had not been completed, but other freight runs had been added to the line. In June 1988 it began handling the Delaware & Hudson Railway's Potomac Yard trains west of Philadelphia. This service had begun in 1976 as a competitive alternative to Conrail, using a combination of Conrail and Amtrak trackage to reach Potomac Yard. The D & H operation started suddenly, following the company's bankruptcy and its temporary operation by the New York, Susquehanna & Western; in the fall of 1989 it was extended past Potomac Yard as a run-through service over the Southern Railway to Linwood, North Carolina. As of mid-1990, the future of the D & H service was questionable but in the meantime the Royal Blue Line was visually spiced by diesels of the D & H, the Susquehanna, the Norfolk Southern lines, and leased units from a variety of owners. In March 1990 a second Conrail train was added on the route.

Two other mid-1980s events also dramatized the Royal Blue Line's personality change. In April 1985 B & O sold the moth-eaten remains of the Staten Island operation to the Delaware-Otsego Corporation, which currently operates it as a short line through its New York, Susquehanna & Western subsidiary. By 1985 the once-strategic railroad was reduced to a single crew working only four or five days a week. The sale meant B & O's complete physical departure from New York, undoubtedly forever. In October of 1989 the NYS & W embargoed four miles of the line between the former Elm Park station and St. George, ending all rail freight service to the once-key terminal.

More positively, in January 1986 CSX opened a transfer terminal for new automobiles at Twin Oaks, Pennsylvania, west of Chester. This facility was another major move in the railroad's strategy to overcome its industrial weakness by building rail-truck intermodal facilities. In this case, Twin Oaks served as both a distribution terminal for new vehicles built in the midwest, and as a loading point for imports. The terminal was later expanded and, by early 1989, was the largest on the East Coast.

Almost as an aside, in 1982 B & O sold its Landenberg branch to Historic Red Clay Valley, Inc., the preservationist group which had used the line for steam excursions since 1966. In 1942 the line had been cut back three miles to Hockessin, Delaware, its current terminal. It now operates under its original name, the Wilmington & Western, primarily as a tourist passenger line — and is scarcely remembered as the seed of the Royal Blue Line.

So the Royal Blue Line passed its centenary alive and healthy, but far from the railroad which John Garrett had ordained and which Daniel Willard had nurtured. In March 1989 B & O's lineal descendant, CSX Transportation, scheduled a total of fifteen freights between Baltimore and Philadelphia, which included two Conrail and two D & H/NYS & W trains. Unquestionably it was a shadow of what it once was, but busy by the standards of the late 1980s. By then virtually all the line's traditional business had disappeared; replacing it were primarily TOFC and auto industry shipments, both markets which barely existed 35 years earlier.

In truth, John Garrett's last railroad project failed in its major mission. However fast and luxurious its passenger trains were, they were never able to overcome the Pennsylvania's head start, its location, and its brute power. And for both passenger and freight business, B & O's presence in New York probably gave it more prestige than profit. By objective standards, the line itself was unnecessary when it was built and redundant for much of its life. But for those very reasons the Royal Blue Line was forced to make itself into something out of the ordinary. That it surely did, and that may be a more lasting accomplishment.

But the memories remain. ***Above:*** *The stodgy but regal Royal Blue, the "civilized" way to New York before the era of Metroliners and air shuttles.* ***Below:*** *The view of New York from the ferry. Train view, B & O Historical Society Collection; ferry photograph by H. H. Harwood, Jr.*

APPENDIX 1

LOCOMOTIVES OF THE ROYAL BLUE LINE

B & O PHILADELPHIA AND NEW YORK DIVISIONS
BALTIMORE BELT LINE ELECTRIC POWER
STATEN ISLAND RAPID TRANSIT

Normally, attempting to compile a complete listing of locomotives used on a single subdivision of a large railroad system is a fruitless and meaningless task. Such rosters tend to be constantly fluid as power is shuttled around the company's divisions and regions. That was true of B & O's Philadelphia line too, but only partially. The Royal Blue Line did have at least three unique categories of motive power: (1) the passenger locomotives specifically designed for it or initially assigned to it; (2) the Baltimore Belt Line's electric "motors," and (3) the isolated and corporately separate Staten Island Rapid Transit — which was unlike anything else on the B & O or, for that matter, anywhere. This appendix presents this more-or-less unique equipment as completely as possible, recognizing the inevitable gaps, particularly in the coverage of freight, switching, and local passenger power.

B & O PHILADELPHIA AND NEW YORK DIVISIONS

Roster Data originally compiled by Howard N. Barr, Sr. and Lawrence W. Sagle

Steam Passenger Locomotives, 1889-1953

Dimensions shown are as originally built.

Numbers	Type	Class	Builder, Date	Driver Diam.	Cylinders	Total Wt. (lbs.)	Tractive Effort (lbs.)	Steam Press.	Notes
838-848	4-4-0	I-5	Baldwin, 1889-90	66"	19" x 24"	106,400	17,299	155	
849-851	4-4-0	M-1	Baldwin, 1890	78"	20" x 24"	114,500	16,738	160	
852-854	4-4-0	M-1	Baldwin, 1891	78"	20" x 24"	114,500	16,738	160	
855-857	4-4-0	M-1/62	Baldwin, 1892	78"	20" x 24"	114,150	17,260	165	
858	4-4-0	M-1/62	Baldwin, 1893	78"	20" x 24"	114,150	17,260	165	
859	4-4-0	M-1 comp	Baldwin, 1893	78"	13.5" & 23" x 24"	123,400	12,883	165	1
1308-1313	4-6-0	B-14	Baldwin, 1896	78"	21" x 26"	145,200	23,740	190	
1319-1322	4-6-0	B-14	Baldwin, 1897	78"	21" x 26"	145,200	23,740	190	
1328-1336	4-6-0	B-17	Baldwin, 1901	78"	15" & 25" x 28"	156,100	21,987	200	2
1424-1449	4-4-2	A-3	Baldwin, 1910	80"	22" x 26"	214,700	27,410	205	
2100-2134	4-6-2	P	ALCo/Schen., 1906	74"	22" x 28"	229,500	32,690	210	3
5100-5129	4-6-2	P-3	Baldwin, 1913	76"	24" x 28"	248,600	34,272	190	
5200-5219	4-6-2	P-5	Baldwin, 1919	73"	25" x 28"	277,000	40,700	200	4
5220-5229	4-6-2	P-5	ALCo/Brooks, 1919	74"	25" x 28"	277,000	40,700	200	4
5300-5319	4-6-2	P-7	Baldwin, 1927	80"	27" x 28"	326,000	50,000	230	5
5320	4-6-2	P-9	B & O Shop, 1928	80"	27" x 28"	329,500	50,000	230	6

Experimental Steam Power, 1935-1937

Numbers	Type	Class	Builder, Date	Driver Diam.	Cylinders	Total Wt. (lbs.)	Tractive Effort (lbs.)	Steam Press.	Notes
2	4-6-4	V-2	B & O Shop, 1935	84"	19" x 28"	294,000	34,000	350	7
5600	4-4-4-4	N-1	B & O Shop, 1937	76"	(4) 18" x 26.5"	391,550	65,000	350	8

Early Diesels, 1925-1938

Numbers	Builder Model	B & O Class	Builder, Date, Construction Numbers	HP	Engine	Weight	Notes
1	—	DE	ALCo/GE/I-R, 1925 #65980/9682	300	6-cyl. I-R 10 x 12	120,000	9
25	HL/2	CG	Plymouth, 1926 #2189	107	gas	36,000	10
50	AA	DE-1	Electro-Motive/GE, 1935 #532/11675	1800	(2) 12-cyl. Winton 201A	252,000	11
51-52	EA	DE-2	Electro-Motive, 1937 #666, 668	1800	(2) 12-cyl. Winton 201A	287,600	12
51-52	EB	DE-2	Electro-Motive, 1937 #667, 669	1800	(2) 12-cyl. Winton 201A	283,000	12
53-56	EA	DE-2	Electro-Motive, 1938 #765-7, 800	1800	(2) 12-cyl. Winton 201A	287,600	12
53-56	EB	DE-2	Electro-Motive, 1938 #768-70, 801	1800	(2) 12-cyl. Winton 201A	283,000	12

NOTES:

1. No. 859 built as Vauclain compound; named *Director General*. Rebuilt 1905 with simple 20" x 24" cylinders and reclassed M-1/62. Held for preservation but scrapped 1942.
2. Nos. 1328-1336 built as Vauclain compounds. Rebuilt 1905 with simple 20" x 28" cylinders. Reclassed B-17a with weight of 148,900 lbs. and tractive effort of 24,410 lbs.
3. Nos. 2100-2134 re# 5000-5034 in 1918; again re# 5150-5184 in 1926.
4. Nos. 5200-5229 were USRA standard light Pacific design.
5. Nos. 5300-5319 originally carried names of United States presidents on cabs. Beginning in 1937 and continuing to 1952, many were modified in various ways. Individual names and major modifications were as follows:

5300	*President Washington*	BLW c.n. 59881	Re# 100 in 1957. Preserved at B & O Museum, Baltimore, Maryland.
5301	*President Adams*	BLW c.n. 59882	To P-7d 1946; re# 109 in 1957.
5302	*President Jefferson*	BLW c.n. 59883	To P-7d 1946; scrapped 1956.
5303	*President Madison*	BLW c.n. 59884	To P-7d 1946; re# 110 in 1957.
5304	*President Monroe*	BLW c.n. 59885	To P-7a for *Royal Blue* 1937; de-streamlined 1939; reclassed P-7. Re-streamlined as P-7d 1946; re# 111 in 1957.
5305	*President Jackson*	BLW c.n. 59886	To P-7c 1944; re# 105 in 1957.
5306	*President Van Buren*	BLW c.n. 59928	To P-7b 1942; re# 104 in 1957.
5307	*President Harrison*	BLW c.n. 59929	Re# 101 in 1957.
5308	*President Tyler*	BLW c.n. 59930	To P-7c 1951; re# 106 in 1957.
5309	*President Polk*	BLW c.n. 59931	To P-7c 1947; re# 107 in 1957. Trailer booster, circa 1933-35.
5310	*President Taylor*	BLW c.n. 59932	Rebuilt as P-9b 1939; rebuilt with conventional firebox 1947; re# 103 in 1957.
5311	*President Fillmore*	BLW c.n. 59933	Scrapped 1953 after overturning.
5312	*President Pierce*	BLW c.n. 59934	To P-7c 1945; to P-7e 1952; re# 112 in 1957.
5313	*President Buchanan*	BLW c.n. 59935	Re# 102 in 1957.
5314	*President Lincoln*	BLW c.n. 59936	To P-7c 1944; re# 113 in 1957. Trailer booster, 1935-42.
5315	*President Johnson*	BLW c.n. 59937	To P-7e 1949; retired 1956.
5316	*President Grant*	BLW c.n. 59973	To P-7e 1952; re# 114 in 1957.
5317	*President Hayes*	BLW c.n. 59974	To P-7c 1945; to P-7e 1952; re# 115 in 1957.
5318	*President Garfield*	BLW c.n. 59975	To P-7c 1944; re# 108 in 1957.
5319	*President Arthur*	BLW c.n. 59976	To P-7e 1951; re# 116 in 1957.

The original olive paint and all striping were replaced with solid blue and names removed from all engines in 1943-44. All P-7 class engines retired 1957-58 except where noted above. Major modifications between 1937 and 1952 were:

P-7a: Streamlined for 1937 *Royal Blue*; cylinders changed to 27.5" x 28"; steam pressure to 240 psi (1938-39). New weight 344,000 lbs.; tractive effort 54,000 lbs. (1938-39). Tender lengthened by 4′ 6".

P-7b: Semi-water tube firebox; type "R" superheater.

P-7c: Solid steel engine beds with cast integral cylinders; pumps on pilot deck; center headlight. Roller bearings added on engine and trailing trucks 1951. Some with semi-water tube firebox.

P-7d: Streamlined for *Cincinnatian*, with some P-7c and P-7e modifications. Roller bearings throughout. New weight 347,000 lbs.; 12-wheel tender.

P-7e: Cast-steel engine beds with integral cylinders; pumps on pilot deck and center headlight; varying applications of roller bearings and modified fireboxes; 12-wheel tender.

P-9b: Water tube firebox replaced conventional firebox.

6. No. 5320 named *President Cleveland*. Built by Mt. Clare shop using boiler from Q-1ba 2-8-2 No. 4201. Originally built with water tube firebox, Caprotti poppet valve gear, and English-style smokebox front. Valve gear replaced with Walschaert in 1929. Water tube firebox improved 1929; reclassed P-9a. Water tube firebox replaced with conventional firebox in 1945 and locomotive reclassed P-7. Retired 1956.

7. No. 2 named *Lord Baltimore*. Trailer booster raised tractive effort to 41,000 lbs. No. 2 transferred to Alton Railroad in 1937 with former *Royal Blue* train set. Re# 5340 and name removed in 1941; returned to B & O in 1942. Stored in 1945 and scrapped 1950.

8. No. 5600 named *George H. Emerson*. Tested on Jersey City run about two years; exhibited at 1939 New York World's Fair; later transferred to service between Washington and Willard, Ohio. Retired 1943 and scrapped 1950.

9. Diesel No. 1 reclassed DS-1 in 1940; re# 195 in 1942; re# 8000 in 1957 and reclassed SA-1. Retired in 1959 and preserved at National Museum of Transport, St. Louis.

10. No. 25 was gasoline-powered with four-wheel truck. Originally worked piers at Philadelphia; later used as shop switcher at Washington, Indiana. Re# 8900 in 1957 and reclassed SP-1; sold 1959.

11. No. 50 transferred to Alton Railroad in 1937 and streamlined front end added. Streamlining removed and used as "B" unit with B & O No. 52 after 1940 (see below). Sold to Alton 1943; re# 100. Rebuilt with six-wheel A-1-A trucks after sale in 1943. Rebuilt with four-wheel trucks 1944. Inherited by GM & O in 1947 and re# 1200. Retired 1957 and preserved at National Museum of Transport, St. Louis.

12. Nos. 51-56 originally operated as paired A-B units carrying one number. Reclassed as DP-2 and DP-2x ("B" units) in 1940. No. 52 transferred to Alton Railroad in 1940 and re# 50 to operate with No. 50 above. No. 52 (50) sold to Alton 1943 and re# 100A. No. 51(A) retired in 1953 and preserved at B & O Museum, Baltimore. Remaining EA-EB units rebuilt as E-8ms (2000 hp) in 1953. "A" units Nos. 53-56 were re# 1434-1437, and "B" units Nos. 51x-56x re# 2414-2419 in 1957.

BALTIMORE BELT LINE ELECTRIC POWER

Roster compiled by Howard N. Barr, Sr.

All locomotives built by General Electric Company.

Numbers	Class	Date Built	Builder Numbers	Weight	Tractive Effort	Driver Diam.	Motors	Notes
1-3	LE-1	1895	1413-5	196,000	49,000	62"	(4) GE-AxB-70	1
5-8	LE-2	1903	1804-7	160,000	40,000	42"	(4) GE-65	2
9	LE-2	1906	2343	160,000	40,000	42"	(4) GE-65	2
11-12	OE-1	1910	Alco 46898-9/ GE 3136-7	184,000	46,000	50"	(4) GE-209	3
13-14	OE-2	1912	Alco 50583-4/ GE 3804-5	200,000	50,000	50"	(4) GE-209	3
15-16	OE-3	1923	Alco 64125-6/ GE 8946-7	242,000	60,500	50"	(4) GE-209A	3
17-18	OE-4	1927	Alco 67225-6/ GE 10341-2	242,000	60,500	50"	(4) GE-209C	3

NOTES:

1. Nos. 1-3 were gearless with four 360 hp motors; frame and body were articulated in two sections. Nos. 1 and 3 retired 1916. No. 2 retired about 1910, later relettered as "No. 1" and exhibited as the "first electric" at various events including the 1927 "Fair of the Iron Horse." It was scrapped about 1935.
2. Nos. 5-9 were boxcabs with all four axles and motors mounted in a single rigid frame, and were designed exclusively for freight service. Nos. 5-8 were delivered as coupled pairs, with No. 9 used as a third unit. No. 5 retired 1917; Nos. 6-9 retired 1934.
3. Nos. 11-18 built by Alco-GE and based on design of Michigan Central (New York Central) Detroit River Tunnel locomotives first built in 1910. Nos. 11-18 re# 151-158 in 1942; re# back to 11-18 in 1948. Generally operated in pairs. Nos. 11-14 upgraded to class OE-3 specifications in 1923. All retired in 1952.

STATEN ISLAND RAPID TRANSIT
STATEN ISLAND RAILWAY

Steam Roster compiled by William D. Edson, Gerald M. Best, and Howard N. Barr

Steam Locomotives, 1860-1946

Staten Island steam power divides into two clear periods, each with its own separate locomotive designs and number series. From the railroad's opening in 1860 to approximately 1906, it used bituminous coal-burning engines — either conventional light 4-4-0s (some purchased new, some second-hand) or Forney design 0-4-4Ts and 2-4-4Ts, similar to elevated railway power of the time. The 0-4-4Ts later became 2-4-4Ts. But as Staten Island became more developed after the turn of the century, the use of anthracite coal became mandatory, requiring new locomotives designed to burn this fuel effectively. So between 1906 and 1910 the existing roster was completely replaced, either with home-built power or newly-purchased camelback locomotives with Wootten fireboxes. In addition, B & O supplied its own power to the SIRT, with some sold outright to its subsidiary, others leased on both long- and short-term assignments. Two classes of B & O camelbacks — D-23 and E-13b — were specifically built or rebuilt for SIRT service and assigned long-term, although always owned by B & O. Various other classes were sent to Staten Island for more temporary use; these included engines from classes B-8, B-18, C-4, D-1, D-3, E-11a, E-14, E-15, E-16a, E-18a, E-19a, E-26, and E-33.

1860-1906:

Number	Type	Builder, Date, Construction Numbers	Driver Diam.	Cylinders	Notes
1	4-4-0	NJ Loco. Works, 3/1860			Named *Albert Journeay;* dropped from roster by 1874.
1 (2nd)	4-4-0	Jersey City Loco. Wks., 1863	60"	16" x 24"	Originally NJRR & T 31, then 131, and finally PRR 731. Acquired circa 1874; gone by 1902.
2	4-4-0	NJ Loco. Works, 3/1860			Named *E. Bancher;* gone by 1884.
3	4-4-0	Danforth Cooke			Probably acquired mid-1860s; gone by 1876.
3 (2nd)	4-4-0	Rogers, 8/1876 #2438	60"	15" x 24"	Bought new; gone by 1902.
4	4-4-0	Danforth, 1/1872 #770	60"	15" x 24"	Bought new; gone by 1902.
5	4-4-0	("Scranton")	60"	17" x 22"	Unknown builder & date; probably from Dickson Works or a DL & W Scranton rebuild. Acquired before 1885; retired 1908.
6	0-4-4T	Cooke, 10/1885 #1641	50"	14" x 22"	Rebuilt as 2-4-4T; re# 15 1892.
6 (2nd)	4-4-0	Baldwin, 8/1892 #12889	60"	16" x 24"	See Note 2
7-9	0-4-4T	Cooke, 10/1885 #1642-4	50"	14" x 22"	Rebuilt as 2-4-4T; see Note 1.
10-14	2-4-4T	Cooke, 2/1887 #1748-52	50"	14" x 22"	See Note 1.
15	2-4-4T	Cooke, 10/1885 #1641	50"	14" x 22"	Ex-1st #6 (1892). Parts used in 2nd #15 1905. See Note 1.
16-18	2-4-4T	Baldwin, 7/1892 #12843-5	50"	14" x 22"	Sold 1906. See Note 1.
45	4-4-0	Rogers, 1866	60"	14" x 20"	Ex-B & O #45; originally Central Ohio #525; acquired by 1886, retired by 1906.

1906-1946:

Numbers	Type	Class	Builder, Date, Construction Numbers	Driver Diam.	Cylinders	Weight	Notes
1	4-4-0	H	Pittsburgh, 8/1889 #1066	60"	18" x 24"	109,200	Ex-B&O #701 H-5; Note 2.
2	4-4-0	H	Pittsburgh, 10/1889 #1084	60"	17" x 24"	109,200	Ex-B&O #792 H-5; Note 2.
3	4-4-0	H	Pittsburgh, 10/1889 #1088	60"	17" x 24"	109,200	Ex-B&O #796 H-5; Note 2.
4	4-4-0	H	Pittsburgh, 11/1889 #1085	60"	18" x 24"	109,200	Ex-B&O #793 H-5; Note 2.
5	4-4-0 (W)	J	Baldwin, 3/1886 #8073	54"	18" x 24"	121,550	Ex-B&O #763 J; Note 3.
6	4-4-0	F	Baldwin, 8/1892 #12889	60"	16" x 24"	81,000	See Note 2.
7-8, 10, 12-22	2-4-4T	R	SIRT Clifton Shop 1905-1910	50"	14" x 22"	100,300	See Note 4.
23-27	4-4-0 (W)	F	ALCo. (Cooke), 8/1906 #40868-72	60"	16" x 24"	116,300	See Note 5.
28	2-8-0 (W)	E	ALCo. (Cooke), 10/1906 #41014	50"	20" x 28"	158,000	Scrapped 1944.
29-30	0-6-0 (W)	D	ALCo. (RI), 2/1908 #44803 4	52"	19" x 28"	161,080	Retired 1944-45.
31-34	4-4-0 (W)	F	ALCo. (Cooke), 5/1910 #47900-3	60"	16" x 24"	116,300	Retired 1927.
B & O 316	0-6-0T	D-1	NJ Loco 1865/B & O Shop 1886, 2/98	44"	19.5" x 22"	97,000	See Note 6.
B & O 1180-4	0-6-0 (W)	D-23	Baldwin, 1/1906 #27171, 83, 96-98	52"	19" x 28"	161,080	Ret. 1944-6; Note 7.
B & O 1630-9	2-8-0 (W)	E-13b	Baldwin, 8/1896, #15014-23; RB 12/04	54"	22" x 28"	173,500	See Note 8.

(W) Indicates Wootten firebox design with center cab (camelback).

NOTES:

1. First Nos. 7-18 were all sold or rebuilt between 1905 and 1908. Parts from Nos. 12, 13, 14, and 15 used in rebuilding second Nos. 12, 13, 14, and 15 in 1905. Others sold to Southern Iron & Equipment Co. (dealer) and resold as follows:
   - 7: To SI & E #687 1909; to Rockwood & Tennessee River #16 7/09; to SI & E #1530 1920; to Greenville & Northern #16 2/20.
   - 8: To SI & E #688 1909; to Dutton Phosphate Co., Gainesville, Florida 5/09.
   - 9: To SI & E #689 1909; to Rockwood & Tennessee River #17, Rockwood, Tennessee 7/09.
   - 10: To SI & E #690 1909; to Enterprise Lumber Co., Goldsboro, North Carolina 8/09; resold to Burr-Holliday Lumber Co., Louise, Mississippi 5/18.
   - 11: To SI & E #691 1909; to Lansing Wheel Barrow Co., Parkin, Arkansas 6/09.
   - 16: To SI & E #543 1906; to Lodwick Lumber Co., Lodwick, Texas 2/07.
   - 17: To Birmingham Rail & Locomotive 1907. Equipped with eight-wheel tender and air brakes; to Arkansas Mill Co., Turkey Creek, Louisiana 5/12; to BR & L; to Scott Lumber Co. 6/17.
   - 18: To SI & E #545 1906; to Kaul Lumber Co., Tuscaloosa, Alabama 1/07.

2. Nos. 1-4 acquired 1902. Nos. 1-4 and 6 from "second roster" were not suitable for anthracite fuel. Nos. 2-4 and 6 retired 1910, sold to Georgia Car & Loco. Works (dealer) and resold as follows:
   - 2: To GC & L #122; to Piedmont & Northern #70 2/13; to GC & L #359 1919; to Bonlee & Western #4, Bonlee, North Carolina 10/19. Scrapped by GC & L 9/24.
   - 3: To GC & L #123; to Pascagoula-Moss Point Northern #36 6/12.
   - 4: To GC & L #124; to Danville & Western #77 1/11; scrapped 1/21.
   - 6: To GC & L #125; to Alabama, Tennessee & Northern #92 3/12.

   No. 1 retired 1912 and scrapped.

3. SIRT #5 (ex-B & O #763) originally built with 69" drivers; refitted with 54" drivers and sold to SIRT 1908. Retired 1920.

4. Nos. 7-8, 10, and 12-22 were built or rebuilt at SIRT's Clifton shop, and had an appearance unique to the SIRT. Their design was based on that of the older narrow-firebox Forney types but with new, higher cylinder saddle, complete new boiler with wide firebox, shorter wheelbase, and newly designed crosshead similar to those used by B & O. These features gave them a shorter, higher look than their Forney predecessors. Certain parts from first Nos. 12, 13, 14, and 15 were used in new locomotives of the same numbers, notably steam dome cover, sandbox, stack, headlight bracket, bell, whistle, and cab fittings. Nos. 9 and 11 were never completed. This group, 7-22 , was used until electrification, and retired between 1927 and 1928. Nos. 13 and 15 sold to Cassaro Construction Co. 1/25 and 2/25; No. 21 sold to Susquehanna River & Western (#6) 9/28.

5. Nos. 23-27 all retired 1927-28, except #25 which was retired 1933.

6. B & O #316 originally built in 1865 as 0-8-0 "Jersey Greenback" #243. Rebuilt 1886 by Mt. Clare Shop to 0-6-0 with new boiler. Converted to tank locomotive and given an overall cab in 1898. Overall cab removed 1943. When retired in 1946 #316 was considered the oldest locomotive in class I railroad service.
7. B & O #1180-84 owned and lettered B & O, but specifically designed and built for SIRT service.
8. B & O #1630-39 originally built 1896 as class E-13 with conventional fireboxes; rebuilt with Wootten fireboxes and center cabs for SIRT service, and reclassed E-13b. Lettered for B & O.

Diesel Locomotives and Electric M.U. Cars, 1925-1971

These rosters cover only equipment owned by SIRT and lettered "Staten Island". Beginning in the late 1960s, the SIRT roster lost its integrity as B & O began moving SIRT diesels to other points on its system and substituting B & O and C & O power on a rotating basis. On July 1, 1971 the New York City Transit Authority took over all SIRT passenger operations and began replacing the original electric car fleet, by then badly depleted by fires, wrecks, and sales.

Diesel Locomotives

Numbers	Builder, Date, Construction Numbers	Builder Model	HP	Weight	Notes
184	General Electric, 9/1943 #18011	65-ton	400	131,100	1
482-484	ALCo.-GE, 12/1943 #71301-3	S-2	1000	230,000	2
485-487	ALCo.-GE, 1/1944 #71309-11	S-2	1000	230,000	2
488-489	ALCo.-GE, 4, 6/1944 #72027-69	S-2	1000	230,000	2

NOTES:

1. No. 184 originally B & O class DS-7g; re# 194 in 1948; re# 8800 and reclassed SG-1 in 1957. Sold 1963 to Manufacturers Equipment Co. (dealer); resold to Pittsburgh Plate Glass, New Martinsville, West Virginia.
2. Nos. 482-489 originally B & O class DS-5a; re# 9026-9033 and reclassed SA-3 in 1957.

Electric Passenger Cars

Numbers	Type	Builder, Date	Notes
300-389	Motor Cars	Standard Steel Car Co., 1925	Class ME-1 after 1928.
390-394	Motor Cars	Standard Steel Car Co., 1925	Rebuilt from trailers 502, 505, 506, 507, and 509 about 1928.
500-509	Trailers	Standard Steel Car Co., 1925	Five cars rebuilt to motors as above. Class MT-1 after 1928.

All cars 67 feet long, 10 feet wide; seating capacity 71. Motor cars had two GE-282 motors and weighed 98,400 lbs.

NOTES:

1. Nos. 319, 358, 376, 378, and 382 destroyed in fire, 1927.
2. Nos. 307, 313, 316, 322, 333, 351, 377, and 390 destroyed in St. George fire, 1946.
3. No. 363 converted to snowplow X-602, 1948, after fire damage in 1946.
4. Nos. 300, 301, 304, 305, 309, 310, 323, 329, 336, 339-342, 344, 345, 349, 350, 354, 357, 362, 364, 373, 383, 387, 391, and 394 (26 cars) retired 1953-1954. Of these, 25 cars were sold to the NYCTA, becoming NYCTA 2900-2924, and used on BMT Culver Line beginning in 1955.
5. Trailers 500, 501, 503, 504, and 508 also sold to NYCTA in 1954 and assigned Nos. 2925-2929, but used only as storage sheds.

APPENDIX 2

ORIGINAL ROYAL BLUE LINE PASSENGER CARS 1890, 1893

Roster compiled by Ralph L. Barger

All cars were built by Pullman Car Works and equipped with narrow enclosed vestibules. Car ownership was split between B & O, Philadelphia & Reading (P & R), and Central Railroad of New Jersey (CNJ) as shown below.

Car Numbers	Owner	Date Built	Builder Lot Number	Comments

Coaches

- Pullman Plan No. 799; B & O Class A-9
- Body length 60 ft.; overall length 67.5 ft.; body width 9.5 ft
- 58 coach seats, 9 seats in smoking compartment: total 67 seats
- Four-wheel trucks

Car Numbers	Owner	Date Built	Builder Lot Number	Comments
1001-1003	CNJ	1890	1737	Re# CNJ 5220-5222 in 1902. Re# 561-63 in 1910. Rebuilt with open platforms before 1910; retired 1931.
1004	CNJ	1893	2012	Re# CNJ 5223 in 1902. Re# 564 in 1910. Rebuilt with open platforms before 1910; retired 1931.
1026-1030	P & R	1890	1733	Re# P & R 1780-1784.
1031	P & R	1893	2012	Re# P & R 1785.
1051-1060	B & O	1890	1730	Re# B & O 5200-5209 in 1901. Rebuilt with wide vestibules and re# 4180-4189 in 1910-11. Nos. 5201-02 rebuilt as coach-baggage cars 1348-9 in 1910.
1061-1063	B & O	1893	2012	Re# B & O 5210-5212 in 1901. Rebuilt with wide vestibules and re# 4190-2 in 1911.

Coach (Smoking)-Baggage Combines

- Pullman Plan No. 824; B & O Class D-5
- Body length 60 ft.; overall length 67.5 ft.; body width 9.5 ft.
- 44 coach seats, 12 seats in smoking compartment: total 56 seats
- Four-wheel trucks

Car Numbers	Owner	Date Built	Builder Lot Number	Comments
1101	CNJ	1890	1738	Re# CNJ 5110 in 1902; re# 199 in 1910. Wrecked 1925.
1111-1112	P & R	1890	1719	Re# P & R 1786-1787.
1121-1123	B & O	1890	1731	Re# B & O 5100-5103 in 1901.
1124	B & O	1893	2013	Re# B & O 5103 in 1901.

Baggage Cars

- Pullman Plan No. 825; B & O Class B-2
- Body length 52 ft. 7 in.; overall length 59 ft. 7.5 in.; body width 9.5 ft.
- Capacity 15 tons
- Six-wheel trucks

Car Numbers	Owner	Date Built	Builder Lot Number	Comments
1151	CNJ	1890	1739	Re# CNJ 5010 in 1902; re# 11 in 1910. Retired 1934.
1152	CNJ	1893	2014	Off roster by 1897; possibly wrecked.
1161-1162	P & R	1890	1734	Re# P & R 1788-1789.
1181-1184	B & O	1890	1732	Re# B & O 5000-5003 in 1901; to 371-4 in 1910.
1185	B & O	1893	2014	Re# B & O 5004 in 1901; to 375 circa 1915.

APPENDIX 3

ROYAL BLUE STREAMLINER CONSISTS, 1935-1958

1935 Lightweight *ROYAL BLUE*

First trip June 24, 1935; removed from New York-Washington service April 1937. After July 1937, became the *Abraham Lincoln* on the Alton Railroad. All cars sold to Alton in 1943 and inherited by GM & O in 1947. Retired in 1969.

- All cars built by American Car & Foundry in 1935 except where noted.

5750	Baggage-mail; rebuilt by B & O shop July 1936 as 44-seat coach-baggage combine.
5800	64-seat coach (smoker); rebuilt by B & O shop July 1935 as bar-buffet-lounge car.
5801	64-seat coach.
5802	64-seat coach.
5806	60-seat coach, built by B & O shop January 1937; steel alloy.
5700	Diner (32 seats)-lunch counter (9 seats).
5930	24-seat parlor car (1 drawing room).
5931	24-seat parlor car (1 drawing room).
5998	18-seat parlor car-lounge-observation.

First 1937 Heavyweight *ROYAL BLUE*

First trip April 25, 1937; after December 9, 1937 renamed the *Columbian*. Removed from New York-Washington service in late 1941 and further rebuilt for Washington-Chicago service beginning December 19, 1941.

- All cars rebuilt by B & O Mt. Clare Shop in 1937 from older cars as shown.

1300	36-seat coach-baggage combine; originally built by Pullman 1926 as coach-baggage combine 1432.
3650	80-seat coach with walkover seats; originally built by Pullman 1926 as coach 5299; rebuilt 1940 as 68-seat coach with reclining seats and renumbered 3520.
3511	54-seat coach-lounge; originally built by Pullman 1927 as coach 5328.
3512	54-seat coach-lounge; originally built by Pullman 1927 as coach 5329.
3065	52-seat coach-buffet; originally built by Pullman 1927 as coach 5326.
1075	Diner (32 seats)-lunch counter (10 seats); originally built by Pullman 1925 as diner 1051.
2110	26-seat parlor car (1 drawing room); originally built by Pullman 1927 as coach 5327.
3300	Buffet-lounge-observation; originally built by Pullman 1927 as coach 5325.

Second 1937 Heavyweight *ROYAL BLUE*

First trip December 9, 1937; remained in service with minor consist changes until April 26, 1958.

- All cars rebuilt by B & O Mt. Clare Shop in 1937 from older equipment as shown.

1301	36-seat coach-baggage combine; originally built by Pullman 1926 as coach-baggage combine 1434.
3521	68-seat coach; originally built by Pullman 1926 as coach 5293.
3513	54-seat coach-lounge; originally built by Pullman 1926 as coach 5291.
3514	54-seat coach-lounge; originally built by Pullman 1926 as coach 5292.
3066	28-seat coach-lunch counter (14 seats); originally built by Pullman 1926 as coach 5295.
1076	Diner (32 seats)-lunch counter (10 seats); originally built by Pullman 1925 as diner 1054. Rebuilt 1938 as full diner with 44 seats.
2111	26-seat parlor car (1 drawing room); originally built by Pullman 1926 as coach 5296.
3301	Buffet-lounge-observation; originally built by Pullman 1926 as coach 5298.

Later consist changes included the addition of un-streamlined 68-seat coach 3528, built in 1923, and the temporary substitution of baggage-buffet car 1317 (ex-Pullman *Capitol Hill*) for combine 1301 between 1953 and 1956.

Data from *Car Names, Numbers and Consists* by Robert J. Wayner and *Equipment Diagrams — The Modernized Heavyweight Royal Blue Trains* published by the Baltimore & Ohio Railroad Historical Society, 1983.

APPENDIX 4

STATIONS OF THE ROYAL BLUE LINE

1. Jersey City-Philadelphia (CRR of NJ and Reading Trackage)

Mileage shown is from Jersey City terminal via Reading's New York Short Line route, used by B & O trains after 1906. The original route via Jenkintown, Pennsylvania added 2.1 miles.

0.0	Jersey City, NJ	CNJ ferry terminal; B & O bus transfer after 1926. Terminal built 1889; rebuilt 1912-14.
11.5	Elizabeth, NJ	CNJ station; stop for all B & O trains. B & O Newark bus connection, 1926-42. Station, built circa 1894.
15.6	Cranford Jct., NJ	Junction with Staten Island Rapid Transit. Not a B & O passenger stop.
23.0	Plainfield, NJ	CNJ station; stop for all B & O trains. WB station, built 1874; EB station built 1901.
31.7	Bound Brook Jct., NJ	CNJ-Reading junction point. Not a B & O passenger stop.
32.8	Manville, NJ	Connection to Lehigh Valley, used by B & O trains to reach Penn Station 1918-26.
68.9	Neshaminy Falls, PA	Junction with New York Short Line route. Not a B & O station stop.
	Newtown Jct., PA	West (south) end of New York Short Line route; junction with ex-North Penn line to Jenkintown and Bethlehem, Pennsylvania.
83.0	Wayne Jct. (Phila.), PA	Reading station; stop for all B & O trains. Transfer point for trains to and from northwest Philadelphia suburban points, Bethlehem, and Allentown, Pennsylvania, and points beyond. Station, built 1901.
88.4	Park Jct. (Phila.), PA	Reading-B & O junction; east end of B & O main line; B & O "PK" tower.
90.0	24th & Chestnut Sta. (Phila.)	B & O Philadelphia passenger station.

2. Philadelphia-Baltimore (B & O Philadelphia Division and Baltimore Belt Line), 1929

This list covers B & O stations and major operating points as they existed in 1929. The locations of interlocking towers and passing sidings varied over the years, but 1929 was arbitrarily chosen as a peak year for stations and facilities. Note that both eastbound and westbound passing sidings were spaced roughly every ten miles to keep freights and local passenger trains clear of through passenger trains.

Station names shown in UPPER CASE letters were originally built as full agency station buildings. The arbitrary letter designations ("A," "B," etc.) indicate the general type of standardized architectural design for each building. These are more fully described at the end of this section. Examples of most designs are illustrated in the text.

Since many stations originally were built at points where no formal communities existed, some station names were changed over the years. The names shown here are those which existed in 1929; earlier and subsequent names are shown in the notes.

All mileages are measured from Park Junction, Philadelphia.

0.0	Park Jct. (Phila.), PA	"PK" tower; junction with Reading's City branch. No station.
1.6	PHILADELPHIA (24th & Chestnut)	Phila. passenger station, built 1886-88; designed by Frank Furness. Demolished 1963.
3.3	East Side (Phila.), PA	"RG" tower; junction with Delaware River branch and entrance to East Side yard. Wooden passenger shelter.
3.7	Eastwick (Phila.), PA	"RJ" tower; junction with Reading's Chester branch; end of Schuylkill River East Side Railroad property.
4.6	58th St. (Phila.), PA	Site of original engine terminal, 1886-1908. No station.
4.7	60th STREET (Phila.), PA	"H" design suburban station, built 1887.

5.0	MT. MORIAH (PHILA.), PA	Wooden two-story station, built 1891, designed by E. F. Baldwin. Discontinued as station before 1919.
6.4	DARBY, PA	"D" design, built 1887.
6.8	BOONE, PA	"H" design suburban station, built 1889. No agent by 1904.
6.9	Boone tunnel	
7.6	COLLINGDALE, PA	"H" design suburban station, built 1887.
8.4	GLENOLDEN, PA	"E" design station, built 1887. Originally named Llanwellyn; renamed to Glenolden in 1920s. WB passing siding between Glenolden and Holmes.
9.5	HOLMES, PA	"E" design, built 1887. Water station. EB passing siding between Holmes and Folsom.
10.5	FOLSOM, PA	"D" design, built 1887.
11.0	RIDLEY PARK, PA	"H" design, built 1887. Originally called Ridley. No agent by 1904.
11.4	Milmont, PA	Wooden shelter.
12.0	EDDYSTONE, PA	Single-story wooden station, built 1889. Originally named Leiperville, then Fairview. Junction with Crum Creek branch; wye.
13.5	CHESTER, PA	Passenger station built 1886; designed by Frank Furness.
14.3	UPLAND, PA	"A" design, built 1889. Previous shelter built 1887. No agent by 1904.
15.7	FELTONVILLE, PA	"G" design, built circa 1888; designed by A. H. Bieler. Originally called Felton. No agent by 1919. EB and WB passing sidings.
17.1	TWIN OAKS, PA	"E" design, built 1887. Originally called Village Green; name changed circa 1890. No agent by 1922.
18.1	BOOTHWYN, PA	"G" design, built 1888; designed by A. H. Bieler.
18.8	OGDEN, PA	Station built 1905; previous shelter built 1887.
19.8	CARPENTER, DE	"E" design, built 1887. No agent by 1922.
21.1	Arden, DE	Shelter, built 1886. Originally called Harvey.
22.0	Silver Side, DE	"SD" tower; EB and WB passing sidings. Shelter station, built 1887.
23.1	CARRCROFT, DE	"D" design, built 1887. Burned and replaced by single-story wooden station 1897. No agent by 1904.
25.4	Concord, DE	Shelter, built 1887.
26.6	WILMINGTON, DE	Passenger station, built 1886-87; designed by Frank Furness. Water station; EB and WB passing sidings.
28.4	Elsmere Jct., DE	"JU" tower; crossing of Reading's W & N branch and junction with B & O Market Street (Wilmington) branch. Reading-owned interlocking tower and station. Originally called W & N Junction.
29.2	Wilsmere yard	Main freight yard for Wilmington area and for through freight movements. Engine terminal and water station. No station.
30.4	Landenberg Jct., DE	"WJ" tower; junction with B & O Landenberg branch. Originally called West Junction.
31.6	KIAMENSI, DE	"C" design, built 1887. No agent by 1929.
32.8	Stanton, DE	Shelter, built 1887. Water station and track pans.
35.1	Harmony, DE	Shelter, built 1886.
38.8	NEWARK, DE	Brick and wooden station, built 1886-87; replaced 1945. Possibly designed by Frank Furness. EB and WB passing sidings.
42.3	Barksdale, MD	Shelter, built 1886.
42.9	ELK MILLS, MD	"A" design, built circa 1889. Originally called Baldwin. Water station.
44.3	SINGERLY, MD	"F" design, built 1886. "SY" tower east of station with EB and WB passing sidings; wye.
45.5	CHILDS, MD	"C" design, built 1887. Junction with B & O Providence Mill ("LC & S") branch after 1892.
50.4	LESLIE, MD	"C" design, built 1887.
53.3	Foy's Hill, MD	"FH" tower; EB and WB passing sidings. No station.
56.1	JACKSON, MD	"F" design, built 1886. Originally called Whitaker. No agent by 1904. "B & O holly tree" located east of station.
57.8	AIKIN, MD	"A"-type design, built 1887; designed by Frank Furness and apparently the prototype for later standardized "A" design. Originally called Frenchtown; named changed 1888 or before. Later spelled "Aiken". "SA" tower west of station; EB and WB passing sidings.
58.3	East end, Susquehanna bridge	

59.4	West end, Susquehanna bridge	
60.2	HAVRE DE GRACE, MD	"F" design, built 1886.
63.0	Swan Creek, MD	Shelter, built 1886. Water station with track pans.
65.0	ABERDEEN, MD	"B" design, built 1886. "A" tower west of station, with EB and WB passing tracks. As of 1990, station still survived as the last original station on the line.
67.2	STEPNEY, MD	"E" design, built 1887. No agent by 1929.
69.8	BELCAMP, MD	Single-story wooden board-and-batten station with large attached freight house, built before 1891. Originally called McGaw's.
70.9	SEWELL, MD	"A" design, built circa 1888. Originally called Harford.
73.5	VAN BIBBER, MD	"F" design, built 1886. Water station. Originally called Preston; renamed by 1888.
75.3	Clayton, MD	Shelter, built 1886.
75.8	"CN" tower	EB and WB passing sidings.
76.6	JOPPA, MD	"C" design, built 1887.
77.9	BRADSHAW, MD	"C" design, built 1887.
79.7	LORELEY, MD	Single-story frame station, built 1910; no agent by 1929. Originally called Morrison.
81.8	COWENTON, MD	"B" design, built 1886. Later called White Marsh.
84.3	Poplar, MD	Shelter, built 1886. "BR" tower; EB and WB passing sidings.
86.0	Rossville, MD	Shelter, built 1886.
88.4	Rosedale, MD	Shelter, built 1886.
89.6	Herring Run, MD	Shelter, built 1886.
90.7	Bay View (Balto.), MD	"BA" tower; junction with Canton branch (original main line) and beginning of Baltimore Belt Line. Freight yard serving industries in area. Water station; wye.
92.8	Gay St. (Balto.), MD	Shelter, built 1892.
93.3	Clifton Park (Balto.), MD	Open platforms for Municipal Stadium, built circa 1922 but seldom used.
94.1	Waverly (Balto.), MD	"SF" tower; east end of Belt Line electrification. Tower built 1901. No station.
95.1	Huntingdon Ave. (Balto.), MD	"HU" tower, built 1897; east end of Belt Line electrification 1895-1901 East end of third and fourth tracks. No station.
95.7	North Ave. (Balto.), MD	"NA" tower; crossing of PRR branch (original Northern Central main line). No station.
96.0	MT. ROYAL (Balto.), MD	Passenger station opened 1896; designed by Baldwin & Pennington. "RM" tower; west end of third and fourth tracks; east portal of Howard Street tunnel.
97.5	CAMDEN STA. (Balto.), MD	Passenger station, originally built 1856-65. West portal of Howard Street tunnel. "DX" tower; water station.

Standardized Station Designs

"A": Very small single-story wood, with decorative dormer. Prototype designed by Frank Furness for Aikin, Maryland (originally Frenchtown) and built of brick and wood. More simplified later all-wooden versions built at Sewell and Elk Mills, Maryland and Upland, Pennsylvania.

"B": 1-1/2 story brick and wood, with attached small freight house. Possibly designed by Frank Furness. Agent's living quarters upstairs. Stations at Aberdeen and Cowenton, Maryland built to this design, but were mirror images of each other.

"C": Single-story brick and wood, with small decorative dormer. Has some Furness characteristics. Built at Bradshaw, Childs, and Leslie, Maryland and Kiamensi, Delaware.

"D": 1-1/2 story brick and wood; agent's living quarters on first floor. Built at Darby and Folsom, Pennsylvania and Carrcroft, Delaware. Some Furness characteristics.

"E": Two-story wooden frame residence-style building, with shed dormers and fish-scale wooden siding on upper portions. Agent's living quarters upstairs. Built at Holmes, Glenolden (Llanwellyn), and Twin Oaks, Pennsylvania; Stepney, Maryland; and Carpenter, Delaware.

"F": 1-1/2 story brick and wood, with high peaked roof, multiple decorative dormers, fish-scale wooden siding, and small attached freight house. Probably a Furness design. Built at Havre de Grace, Jackson (Whitaker), Singerly, and Van Bibber, Maryland. Havre de Grace and Van Bibber stations apparently were mirror images of Jackson and Singerly.

"G": Single-story wooden frame, with decorative eyelid dormer and rounded agent's bay. Designed by Baltimore architect A. H. Bieler.

"H": Single-story brick suburban design with wooden decorative trim and wooden *porte-cochere*. Eyelid dormer. Some Furness characteristics. Built at 60th Street (Philadelphia), Boone, Collingdale, and Ridley, Pennsylvania.

3. B & O LANDENBERG BRANCH

Former Wilmington & Western (Delaware Western) main line, originally built 1871-1872. Most existing stations were single-story flat-roofed wooden structures of a standardized design unique to the W & W, and apparently were built at the time the railroad itself was built.

Station names shown in UPPER CASE letters were originally agency stations. Mileages shown are from Landenberg Junction, Delaware.

0.4	MARSHALLTON, DE	
1.0	GREENBANK, DE	No agent by 1904.
1.6	Brandywine Springs, DE	
1.9	Faulkland, DE	
3.3	WOODDALE, DE	
5.0	Mt. Cuba, DE	
6.5	ASHLAND, DE	
7.7	YORKLYN, DE	Station later relocated to Greenbank, Delaware by Wilmington & Western excursion rail road. Last surviving original station and typical of the "standard" early W & W design.
9.6	HOCKESSIN, DE	
11.2	SOUTHWOOD, DE	
12.0	Eden, PA	
12.5	Broad Run, PA	
14.3	LANDENBERG, PA	Junction with PRR Pomeroy branch.

4. STATEN ISLAND RAPID TRANSIT-STATEN ISLAND RAILWAY

All mileages shown from St. George, New York. All stations had passenger facilities except where noted.

St. George-Tottenville Line (all stations in New York)

0.0	St. George	Passenger ferry terminal and freight marine terminal.
0.6	Tompkinsville	
1.3	Stapleton	
1.7	Clifton	Junction with South Beach branch; location of SIRT shops. East (north) end of SI Railway property and original terminal of SI Railway.
3.2	Grasmere	
4.5	Dongan Hills	
5.4	Grant City	
5.9	New Dorp	
6.8	Oakwood Heights	
7.5	Bay Terrace	Earlier station named Whitlock slightly east, largely abandoned by 1912.
8.3	Great Kills	
9.2	Eltingville	
9.9	Annadale	
10.9	Huguenot Park	
11.6	Princess Bay	
12.5	Pleasant Plains	
13.1	Richmond Valley	
14.0	Atlantic	
14.4	Tottenville	

South Beach Branch (all stations in New York)

1.7	Clifton	Junction with St. George-Tottenville line
2.2	Rosebank	
2.7	Fort Wadsworth	
3.1	Arrochar	
4.1	South Beach	

St. George-Arlington-Cranford Jct. Line (stations in New York except where noted)

0.0	St. George	
0.7	New Brighton	
1.3	Sailors Snug Harbor	
1.8	Livingston	
2.4	West New Brighton	
3.0	Port Richmond	
3.4	Tower Hill	
3.9	Elm Park	
4.6	Mariners Harbor	
5.2	Arlington	West terminal of passenger services.
6.9	Arthur Kill bridge	
8.0	Bayway, NJ	
8.7	Linden Jct., NJ	Freight interchange with PRR.
11.8	Staten Island Jct., NJ	Freight interchange with LVRR.
12.2	Cranford Jct., NJ	Freight interchange with CRR of NJ. Water station.

Bibliography

1. References covering the B & O portion of the Royal Blue Line, B & O-PRR relationships, and general background:

Allodi, E. F. "Architects Had an Engaging Problem in the Design of Our New Forty-Second Street Motor Coach Terminal in New York." *Baltimore & Ohio Magazine* (February 1929).

American Railway Guide and Pocket Companion, 1851. New York, New York: Curran Dinsmore & Company, 1851; reprinted by Kalmbach Publishing Company, Milwaukee, Wisconsin, 1945.

"The Baltimore and Ohio Navy." *Baltimore and Ohio Employes [sic] Magazine* (October 1916).

Baltimore & Ohio Railroad. *Annual Reports, 1884-1959.*

"The Baltimore & Ohio Railroad Company and its Passenger and Freight Interests in New York City." *Book of the Royal Blue* (September 1907).

"Baltimore & Ohio Railroad Terminal at Philadelphia." *Book of the Royal Blue* (November 1906).

Baltimore & Ohio Railroad Company *Corporate Histories — Vol. 2.* Baltimore, Maryland: Baltimore & Ohio Railroad Valuation Department, 1922.

Baltimore & Ohio Railroad Historical Society. *Equipment Diagrams — The Modernized Heavyweight Royal Blue Trains.* Baltimore, Maryland: Baltimore & Ohio Railroad Historical Society, 1983.

"B & O Buses." *Motor Coach Age* (July-August 1971).

"B & O's Two New Lightweight, High Speed Trains Ready for Service." *Baltimore & Ohio Magazine* (June 1935).

Barr, Howard N., Sr. "The Lord and the Lady." *B & O Railroader* (January-February 1976). (Treats the development and operation of the *Lady* and *Lord Baltimore* locomotives.)

Bendersky, Jay. *Brooklyn's Waterfront Railways.* East Meadow, New York: Meatball Productions, 1988. (Covers Brooklyn terminal lines which acted as B & O agents, and also the phase-out of harbor operations in the 1970s.)

Bezilla, Michael. *Electric Traction on the Pennsylvania Railroad, 1895-1968.* University Park, Pennsylvania: The Pennsylvania State University Press, 1980.

Burgess, George H., and Miles C. Kennedy. *Centennial History of the Pennsylvania Railroad Company.* Philadelphia, Pennsylvania: The Pennsylvania Railroad Company, 1949.

Catton, William Bruce. "John W. Garrett of the Baltimore & Ohio: A Study in Seaport and Railroad Competition, 1820-1874." Ph.D. dissertation, Northwestern University, 1959.

Condit, Carl W. *The Port of New York* (two volumes). Chicago, Illinois: University of Chicago Press, 1980, 1981.

Cornell, W. "Organizing a Lighterage Department." *Baltimore and Ohio Employes Magazine* (May 1914). (Describes early B & O New York harbor operations.)

"From New York to Baltimore and Washington by Rail." *Engineering News* (February 27, 1892).

Gibb, Hugh R. "Brotherly Love — Philadelphia Style (Wherein the Baltimore & Ohio Attempts to Reach New York)." *National Railway Historical Society Bulletin* (Vol. 39, No. 6, 1974).

Harwood, Herbert H., Jr. *Impossible Challenge: The Baltimore & Ohio Railroad in Maryland.* Baltimore, Maryland: Barnard, Roberts & Company, 1979.

Howes, William F. "The Royal Blue." *The Railroad Enthusiast* (Vol. 14, Nos. 1-2, 1977).

Hungerford, Edward. *The Story of the Baltimore & Ohio Railroad* (two volumes). New York, New York: G. P. Putnam's Sons, 1928.

Keyser, William. "Recollections of a Busy Life." Unpublished typescript, circa 1901, in collection of Maryland Historical Society, Baltimore, Maryland.

Kirkland, John F. *Dawn of the Diesel Age: The History of the Diesel Locomotive in America.* Glendale, California: Interurban Press, 1983.

Klein, Maury. *The Life and Legend of Jay Gould.* Baltimore, Maryland: The Johns Hopkins University Press, 1986.

Kline, Benjamin F. G. *Little, Old & Slow: The Life and Trials of the Peach Bottom and L.O. & S. Railroads.* Lancaster, Pennsylvania: Benjamin F. G. Kline, 1985. (Gives history of Lancaster, Cecil & Southern branch.)

Kuhler, Otto. *My Iron Journey: A Life with Steam and Steel.* Denver, Colorado: Intermountain Chapter, National Railway Historical Society, 1967.

Locomotive Engineering (May 1891). (Royal Blue Line locomotives and services.)

"New Diesel-electrics Making History on B & O." *Baltimore & Ohio Magazine* (June 1937).

"New Forty-second Street Passenger Station Opened in New York." *Baltimore & Ohio Magazine* (January 1929).

"The New Gateway to the Metropolis." *Book of the Royal Blue* (December 1905). (Describes 23rd Street ferry terminal.)

"New Motor Coach Train Connection Station in Rockefeller Center." *Baltimore & Ohio Magazine* (June 1937).

"The New 23rd Street New York Terminal of the Baltimore & Ohio Railroad." *Book of the Royal Blue* (October 1905).

Official Guide of the Railways and Steam Navigation Lines of the United States, various monthly issues, 1869-1958. (Title varies; in earlier years titled *Travellers' Official Railway Guide* with varying subtitles.)

O'Gorman, James. *The Architecture of Frank Furness.* Philadelphia, Pennsylvania: Philadelphia Museum of Art, 1973. (Describes Furness-designed stations at Philadelphia and Chester, Pennsylvania.)

"Our Latest Family Connection." *Baltimore & Ohio Magazine* (September 1926). (Covers initial New York motor coach operations.)

"Passenger Train Service Between Washington and New York." *Railway Age* (April 5, 1895). (Describes high-speed operations on Royal Blue Line route.)

Pawson, John R. *Delaware Valley Rails.* Willow Grove, Pennsylvania: John R. Pawson, 1979. (Describes B & O facilities in the Philadelphia area.)

Pennypacker, Bert. "A Royal Blue Line Recall." *Trains* (August 1972).

"Rapid Bridge Building." *Book of the Royal Blue* (May 1906). (Bridge replacement at Swan Creek, Maryland.)

Rosenbaum, Joel, and Tom Gallo. *Iron Horses Across the Garden State: New York to Philadelphia by Rail.* Piscataway, New Jersey: Railpace Company, 1985.

"The Royal Blue Gets off to a Royal Start." *Baltimore & Ohio Magazine* (July 1935).

"The Royal Limited." *Book of the Royal Blue* (December 1898).

Sagle, Lawrence W. *B & O Power.* Medina, Ohio: Alvin F. Staufer, 1964.

Sagle, Lawrence W. "Thomas Leiper's Railroad." *Trains* (February 1943). (History of B & O's Crum Creek branch.)

"Spring Schedules Bring New Trains." *Baltimore & Ohio Magazine* (April 1937). (Describes "Improved" *Royal Blue.*)

Stover, John F. *History of the Baltimore & Ohio Railroad.* West Lafayette, Indiana: Purdue University Press, 1987.

Thompson, Everett L. "Definition of a Redball." *Trains* (June 1966). (Describes operation of B & O merchandise trains 117-118 in 1938.)

"Track Tanks." *Book of the Royal Blue* (July 1909).

Volkman, Arthur G. *The Story of the Wilmington & Western Railroad.* Wilmington, Delaware: Historic Red Clay Valley, Inc., 1963.

Ward, James A. *J. Edgar Thomson: Master of the Pennsylvania.* Westport, Connecticut: Greenwood Press, 1980.

Warner, Paul T. "The Washington-New York Service of the Baltimore & Ohio Railroad." *Baldwin Locomotives* (October 1927).

Wayner, Robert J. *Car Names, Numbers and Consists.* New York, New York: Wayner Publications, 1972.

White, John H., Jr. *The American Railroad Passenger Car.* Baltimore, Maryland: The Johns Hopkins University Press, 1978.

Williams, Earl P., Jr. *The Architecture and Engineering of Amtrak's Washington-New York Corridor: A Pictorial History.* Lanham, Maryland: Maryland Historical Press, 1977. (Data on the Pennsylvania Railroad's Baltimore tunnels.)

Wilson, William Bender. *History of the Pennsylvania Railroad Company: With Plan of Organization, Portraits of Officials and Biographical Sketches.* Philadelphia, Pennsylvania: Henry T. Coates & Company, 1899.

Yuhas, H. Michael. "B & O's Philadelphia Subdivision." *Railpace Newsmagazine* (July and August, 1986).

2. Specific references on the Baltimore Belt Railroad and Howard Street tunnel:

"The Baltimore Belt Railroad." *Engineering News* (issues of December 12 and December 19, 1891). (Detailed description of tunnel construction techniques.)

"The Baltimore and Ohio Third-Rail System." *Railway Age* (July 25, 1902).

Barr, Howard N., Sr. "The Baltimore Belt Line." Unpublished manuscript, 1974.

Barr, Howard N., Sr. "Triple Level Crossing." *The Sentinel*, B & O RR Historical Society (May-June 1987). (Details of North Avenue bridges and tunnels.)

Duke, Donald. "America's First Main Line Electrification." In *American Railroad Journal*. San Marino, California: Golden West Books, 1966.

"Electric Freight Locomotive." *Canadian Electrical News and Steam Engineering Journal* (January 1896). (Describes experimental freight operations and testing.)

Hilton, George W. *Ma & Pa: A History of the Maryland & Pennsylvania Railroad*. Berkeley, California: Howell-North Books, 1963; republished 1980. (Covers Maryland Central's early involvement in the Belt project.)

Middleton, William D. *When the Steam Railroads Electrified*. Milwaukee, Wisconsin: Kalmbach Publishing Company, 1974.

"Operation of the Baltimore and Ohio Electrification." *Electric Railway Journal* (June 10, 1916).

Railroad Gazette (various dates, 1894-1901). Feature articles of particular interest are as follows:

December 14, 1894: "The Baltimore Belt Railroad."

January 11, 1895: "Portals of the Howard Street Tunnel, Baltimore."

May 10, 1895: "The Baltimore Belt Line Terminal." (Mt. Royal station).

June 28, 1895: "Baltimore Belt Line Power Station and Equipment."

July 19, 1895: "Electrical Equipment of the Baltimore Belt Line Tunnel."

October 4, 1895: "The Baltimore Tunnel Electric Locomotive in Service."

November 8, 1895: "B and O Electric Locomotive."

March 6, 1896: "Testing the B and O Electric Locomotive."

May 15, 1896: "The New Baltimore and Ohio Station in Baltimore." (Mt. Royal).

October 15, 1897: "Recent Improvements on the Baltimore and Ohio: Terminal Improvements in Baltimore."

January 11, 1901: "Electrical Equipment of the B and O Tunnel."

Sagle, Lawrence W. "Baltimore & Ohio Stations in Baltimore." Railway & Locomotive Historical Society *Bulletin 106* (1962).

Sagle, Lawrence W. "Baltimore Belt Line." *Trains* (April 1943).

3. References covering history and operations of the Philadelphia & Reading (Reading Company) and Central Railroad of New Jersey as they relate to the Royal Blue Line:

Anderson, Elaine. *The Central Railroad of New Jersey's First 100 Years: A Historical Survey*. Easton, Pennsylvania: Center for Canal History & Technology, 1984.

Bogen, Jules I. *The Anthracite Railroads: A Study in American Enterprise.* New York, New York: Ronald Press Company, 1927.

Crater, Warren B. *Locomotives of the Jersey Central, 1-999* (Revised Edition). Roselle Park, New Jersey: Railroadians of America, 1978.

Hare, Jay V. *History of the North Pennsylvania Railroad.* Philadelphia, Pennsylvania: Reading Company, 1944.

Hare, Jay V. *History of the Reading*. Philadelphia, Pennsylvania: John H. Strock, 1966. (Republication of articles serialized in *The Pilot*, P & R employees' magazine, 1909-1914.)

"Jersey Central Engine Terminal at Communipaw." *Railway Age Gazette* (June 26, 1914).

"The Jersey City Passenger Station Improvements." *Railway Age Gazette* (November 6, 1914).

"The Manhattan Terminal of the Central Railroad of New Jersey." *Railway Age Gazette* (November 6, 1908).

"New Station and Interlocking at the Jersey City Terminus of the Central of New Jersey." *Railroad Gazette* (June 28 and October 18, 1889).

"Reconstruction of the Jersey City Terminal Yards." *Railway Age Gazette* (April 9, 1915).

"Remodeling Jersey City Passenger Station." *Engineering Record* (June 27, 1914).

Wiswesser, Edward H. *Steam Locomotives of the Reading and P & R Railroads*. Sykesville, Maryland: Greenberg Publishing Company, 1988.

4. References covering the Staten Island Rapid Transit:

"The Arthur Kill Bridge." *Railroad Gazette* (June 22, 1888).

"The Arthur Kill Bridge Approaches." *Railroad Gazette* (July 26, 1889).

Bogart, Stephen. "Little Known Railroad." *Trains* (February 1951).

Cornell, W. "Recollections of Early Railroading on Staten Island." *The Baltimore and Ohio Employes Magazine* (May 1914).

Dryden, W. L. "Progress of Signalling on Staten Island Lines." *The Baltimore and Ohio Employes Magazine* (May 1914).

Glucksman, Randy. "Progress Across the Bay." *ERA Headlights* (May-June 1972). (Covers sale of line to NYCTA.)

Hilton, George W. *The Staten Island Ferry.* Berkeley, California: Howell-North Books, 1964.

"History of the Baltimore and New York Railway Company." *The Baltimore and Ohio Employes Magazine* (May 1914).

"How the World's Longest Vertical Lift Bridge Will Work." *Engineering News Record* (June 11, 1959).

Krampf, Melvin. "Island Electric." *Railroad Magazine* (July 1949).

Kretzer, S. P. "The Story of Railroad Development on Staten Island." *The Baltimore and Ohio Employes Magazine* (May 1914).

"Representative Industries on Staten Island." *The Baltimore and Ohio Employes Magazine* (May 1914).

Silver Leaf Rapid Transit. *Staten Island Rapid Transit.* Brooklyn, New York: Silver Leaf Rapid Transit, 1965.

"Staten Island Electrification." *Railway Age* (July 14 and November 21, 1925).

Travis, Rich. "From Chessie to Susie-Q." *Railpace Newsmagazine* (March 1986). (Covers sale of SIRT to Delaware-Otsego and contemporary operations.)

INDEX

DEDICATION

To the memory of
Charles H. Detwiller (1863-1940)
who, perhaps unknowingly, ignited the first spark.

WITH THANKS TO . . .

Considering that this book covers a comparatively short and simple segment of the Baltimore & Ohio system — and considering that the author has had about fifty years to think about it — the job of writing the story should be quick and easy. Not so at all. In the history of any major enterprise lurk all sorts of complications and unknowns — necessary but undiscovered facts, "hard facts" which turn out to be baseless legends, unknown motives and causes, incomplete documentation, and all manner of unexpected hurdles. But happily, during my fifty years of sitting and thinking, many others had been doing the hard work of finding the facts, collecting the materials, recording the images, and accumulating memories. And more happily, they were uniformly willing to share their work and materials, sometimes at considerable inconvenience.

It is difficult and unfair to try to rank these contributors in any order, since all of them gave whatever they had which related to the subject. In some cases this might have been one or two needed photographs; in others it might have been a lifetime of research. But some kind of "Above and Beyond the Call of Duty Award" must go to **Christopher T. Baer,** Assistant Curator of Manuscripts and Archives at the Hagley Library in Wilmington, Delaware. Besides making available the pricelessly helpful documents in the Hagley Collection, Chris shared his own bottomless knowledge of all facets of eastern railroad history and his personal research work on the Jersey Central, Reading, and B & O. In addition, he meticulously reviewed, corrected, and supplemented the original manuscript. Another major support was **Gary W. Schlerf,** president of the B & O Railroad Historical Society, who made available numerous photographs, drawings, and company documents from the Society's large archives and his own collection.

There is no source like firsthand working experience, and I was fortunate in getting it from two viewpoints. **Jervis Langdon, Jr.,** B & O's president between 1961 and 1964, contributed many of the facts and most of the insights into this critical period in the company's history. And the late **Everett L. "Tommy" Thompson,** who spent much of his professional career in B & O's passenger department, patiently explained how and why things happened as they did during the last three decades of the New York services. Tommy also loaned his comprehensively detailed on-board logs of many of his train rides over the line during this period. Sadly, he did not live to see the finished product, but he is very strong in my memory.

In addition to material from their own collections, **Frank A. Wrabel** and **Joseph J. Snyder** both supplied their professional expertise and creativity in suggestions on the book's design and the mechanics of presentation.

Credit also must be given to **Hugh R. Gibb** and **Howard N. Barr, Sr.** for their original research and writing on specific aspects of this story. Hugh Gibb struggled through masses of original source materials to write a comprehensive and revealing history of the B & O's problems in building into Philadelphia. Howard Barr did much the same in documenting the Baltimore Belt Line's construction, facilities, operations, and equipment. Howard is unquestionably the foremost living expert on B & O locomotives and was invaluable in helping with the complex details of the motive power rosters and in providing much supplementary information. **Michael F. Kotowski,** a superbly accomplished California artist, was responsible for the dramatic cover painting.

Others who helped with data, materials, manuscript review, and general encouragement were:

Carlos P. Avery
Ralph L. Barger
Mark S. Bennett
Robert W. Breiner
Richard J. Cook
Milton A. Davis
James D. Dilts
Donald Duke
William D. Edson
Thomas R. Flagg
James P. Gallagher
John P. Hankey
John C. Hayman
E. Henry Hinrichs
Andrew A. Holzopfel
Harry Jones
Robert R. Malinoski
Ara Mesrobian
Rodney H. Peterson
Nicolas C. Powell
Donald A. Somerville
Raymond T. Stern
Preston Thayer
Robert J. Wayner
Frank A. Weer
John H. White, Jr.
William E. Worthington, Jr.
Robert M. Vogel
Theodore A. Xaras
Robert J. Yanosey

And finally, the professional staff at Greenberg Publishing: **Terri Glaser** who made me semi-computer literate and as copy editor, straightened out my English; **Marcy Damon** who made a finished book out of a jumble; **Maureen Crum** who produced comprehensible maps where none existed; **Donna Price** who proofread the text and sized the many photographs; and **Samuel Baum** who solved all the problems which no one else could.

This is a long list of impressive names, and I am grateful to every one of them. Without them I would be another fifty years at this, but without much to show for it all.

Herbert H. Harwood, Jr.
June 1990

THE AUTHOR

Herbert H. Harwood, Jr. has combined a working railroad career with a lifelong interest in railroading, perhaps the result of having a father and a grandfather in the business. After receiving academic degrees in history and in transportation economics, he started in the Chesapeake & Ohio's finance department in Cleveland. Later he moved into marketing for C & O, the Chessie System, and their latest successor, CSX Transportation. Along the way he has written numerous books and articles on many phases of railroad history, including *Impossible Challenge: The B & O in Maryland.* Now retired, Herb lives in Baltimore with his wife, Janice, an art historian and teacher.

THE COLUMBIAN
BALTIMORE & OHIO
NEW YORK-WASHINGTON